D1433878

MATHS
in
BITE-SIZED
CHUNKS

Also by the same author:

I Used to Know That: Maths
From 0 to Infinity in 26 Centuries

MATHS
in
BITE-SIZED
CHUNKS

CHRIS WARING

First published in Great Britain in 2018
by Michael O'Mara Books Limited
9 Lion Yard
Tremadoc Road
London SW4 7NQ

A CIP catalogue record for this book is available from the British Library.

Papers used by Michael O'Mara Books Limited are natural,
recyclable products made from wood grown in sustainable forests. The
manufacturing processes conform to the environmental regulations of
the country of origin.

ISBN: 978-1-78243-846-5 in hardback print format
ISBN: 978-1-78243-848-9 in ebook format

1 2 3 4 5 6 7 8 9 10

www.mombooks.com

Designed and typeset by Design 23
Cover design by Dan Mogford

Printed and bound by CPI Group (UK) Ltd, Croydon, CR0 4YY

CONTENTS

4 GEOMETRY

5 STATISTICS

6 PROBABILITY

INTRODUCTION

I could start this book by telling you that maths is everywhere and yammer on about how important it is. This is true, but I suspect you've heard that one before and it's probably not the reason you picked this book up in the first place.

I could start by saying that being numerate and good at mathematics is an enormous advantage in the job market, particularly as technology plays an increasingly dominant role in our lives. There are great careers out there for mathematically minded people, but, to be honest, this book isn't going to get you a job.

I want to start by telling you that skill in mathematics can be learnt. Many of us have mathematical anxiety. This is like a disease, since we pick it up from other people who have been infected. Parents, friends and even teachers are all possible vectors, making us feel that mathematics is only for a select group of people who are just lucky, who were born with the right brain. They do mathematics without any effort and generally make the rest of us feel stupid.

This is not true.

Anyone can learn mathematics if they want to. Yes, it takes time and effort, like any skill. Yes, some people learn it faster than others, but that's true of most things worth learning. I know you're busy, so the premise here is that you want some easily digestible snippets. You can learn them piecemeal, each building on the one before, so that without too much effort you can take on board the concepts that really do explain the world around us.

I've divided the book up into several sections. You'll remember doing a lot of the more basic stuff at school, but my aim is to cover this at a brisk pace to get to the really tasty bits of mathematics that maybe you didn't see. You can work through the book from start to finish, or dip in and out as and when the mood takes you – a six-course meal and a buffet at the same time!

I've also included lots of anecdotes to spice things up – stories of how discoveries were made, who discovered them and what went wrong along the way. As well as being interesting and entertaining, these serve to remind us that mathematics is a field with a vibrant history that tells us a lot about how our predecessors approached life. It also shows that the famous, genius mathematicians had to work hard to get where they got, just like we do.

Prepare yourself for a feast. I hope you're hungry.

1

NUMBER

Chapter 1

TYPES OF NUMBER

Sixty-four per cent of people have access to a supercomputer.

In 2017, according to forecasts, global mobile phone ownership was set to reach 4.8 billion people, with world population hitting 7.5 billion. As the Japanese American physicist Michio Kaku (b. 1947) put it: 'Today, your cell phone has more computer power than all of NASA back in 1969, when it placed two astronauts on the moon.'

At a swipe, each of us can do any arithmetic we need on our phones – so why bother to learn arithmetic in the first place?

It's because if you can perform arithmetic, you start to understand how numbers work. The study of how numbers work used to be called arithmetic, but nowadays we use this word to refer to performing calculations. Instead, mathematicians who study the nature of numbers are called number theorists and they strive to understand the mathematical underpinnings of our universe and the nature of infinity.

Hefty stuff.

I'd like to start by taking you on a trip to the zoo.

Humans first started counting *things*, starting with one thing and counting up in whole numbers (or *integers*). These numbers are called the *natural* numbers. If I were to put these numbers into a mathematical zoo with an infinite number of enclosures, we'd need an enclosure for each one:

$$1, 2, 3, 4, 5, 6 \ldots$$

The ancient Greeks felt that zero was not natural as you couldn't have a pile of zero apples, but we allow zero into the natural numbers as it bridges the gap into *negative* integers – minus numbers. If I add zero and the negative integers to my zoo, it will look like this:

$$\ldots -6, -5, -4, -3, -2, -1, 0, 1, 2, 3, 4, 5, 6 \ldots$$

My zoo now contains all the negative integers, which when combined with the natural numbers make up the group of numbers called, imaginatively, the *integers*. As each positive integer matches a negative one, my zoo needs twice as many enclosures as before, with one extra room for zero. However, my infinite mathematical zoo does not need to expand, as it is already infinite. This is an example of the hefty stuff I referred to earlier.

There are other types of numbers that are not integers. The Greeks were happy with piles of apples, but we know an apple can be divided and shared among a number of people. Each person gets a fraction of the apple and I'd like to have an example of each fraction in my zoo.

If I want to list all the fractions between zero and one, it would make sense to start with halves, then thirds, then

Fractions

Fractions show numbers that are between whole integers and are written as one number (the numerator) above another (the denominator) separated by a fraction bar. For example, a half looks like:

$$\frac{1}{2}$$

One is the numerator, two is the denominator. The reason it is written this way is that its value is one divided by two. It tells you what fraction of something you get if you share one thing between two people. $\frac{3}{4}$ is three things shared between four people – each person gets three quarters.

quarters, etc. This methodical approach should ensure I get all the fractions without missing any. So, you can see that I'm going to have to go through all the natural numbers as denominators (the numbers on the bottom of the fraction). For each different denominator, I'll need all the different numerators (the numbers on the top of the fraction), starting from one and going up to the value of the denominator.

Once I've worked out all the fractions between zero and one, I can use this to fill in all the fractions between all the natural numbers. If I add one to all the fractions between zero and one, this will give me all the fractions between one and two. If I add one to all of them, I'll have all the fractions between three and four. I can do this to fill in the fractions between all the natural numbers, and I could subtract to fill in all the fractions between the negative integers too.

So, I have infinity integers and I now need to build infinity enclosures between each of them for the fractions. That means I need infinity times infinity enclosures altogether. Sounds like a big job, but luckily I still have enough enclosures.

As the fractions can all be written as a ratio as well, the fractions are called the *rational* numbers. I now have all the rational numbers, which contain the integers (as integers can be written as fractions by dividing them by one), which contain the natural numbers in the zoo. Finished.

Just a moment – some mathematicians from India 2,500 years ago are saying that there are some numbers that can't be written as fractions. And when they say 'some', they actually mean infinity. They discovered that there is no number that you can square (multiply by itself) to get two, so the square root of two is not a rational number. We can't actually write down the square root of two as a number without rounding it, so we just show what we did to two by using the *radix* symbol: $\sqrt{2}$. There are other really important numbers that are not rational that have been given symbols instead as it is a bit of a faff to write down an unwritedownable number: π, e and φ are three examples that we'll look at later. We call such numbers *irrational*, and I need to put these into the zoo as well. Guess how many irrational numbers there are between consecutive rational numbers? That's right – infinity! However, I can still squeeze these into my infinite zoo without having to build any more enclosures, although Cantor might have a thing or two to say about that (see page 17).

Squares and Square Roots

When you multiply a number by itself, we say the number has been *squared*. We show this with a little two called a power or index:

$$3 \times 3 = 3^2$$

Three squared is nine. This makes three the *square root* of nine. Square rooting is the opposite of squaring. The square root of sixteen is four because four squared is sixteen:

$$\sqrt{16} = 4$$

Numbers like nine and sixteen are called *perfect* squares, because their square root is an integer. Any number, including fractions and decimals, can be squared. Any positive number can be square rooted.

For much more information about this, see page 58.

When we put the irrational numbers together with the rational numbers we have what mathematicians call the *real* numbers. If you've spotted a pattern in what went before, you'll suspect that there are also not-real numbers and you'd be right. However, I'm going to stop there and name my zoo *The Infinite Real Number Zoo*. Most zoos sort their animals out by type, so I could organize mine into overlapping groups of types. The map might look like this, and I've put a few must-sees in to help you plan your day out:

The Infinite Real Number Zoo

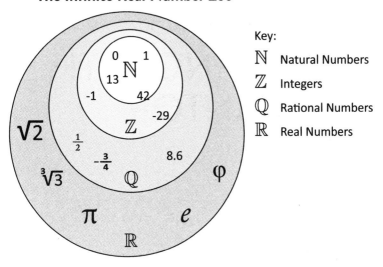

Key:

$\mathbb{N}$ Natural Numbers

$\mathbb{Z}$ Integers

$\mathbb{Q}$ Rational Numbers

$\mathbb{R}$ Real Numbers

I must own up to the fact that my zoo owes a lot to the German mathematician David Hilbert (1862–1943). He made great contributions to mathematics but is best known for his advocacy and leadership of the subject. In 1900 he produced a list of twenty-three unsolved problems – now known as the Hilbert problems – for the International Congress of Mathematicians, three of which are still unresolved to this day. The thought experiment *Hilbert's Hotel*, the source for my zoo, concerns Hilbert's musings on a hotel with an infinite number of rooms filled with an infinite number of guests. Hilbert shows that we can still fit another infinite number of guests into the hotel if we can persuade all the initial guests to move to the room with a number double their current room number. The current guests would all now be in even-numbered rooms, leaving the odd-numbered rooms (of which there are infinitely many) for the new arrivals.

Chapter 2

COUNTING WITH CANTOR

Galileo Galilei (1564–1642) came up with a nice puzzle known as *Galileo's paradox* while under house arrest in Italy for his heretical belief that the earth went around the sun.

It says that while some natural numbers are perfect squares (see page 15), most are not, so there must be more not-squares than squares. However, every natural number can be squared to produce a perfect square, so there must be the same number of squares as natural numbers. Hence, a paradox: two logical statements that cannot both be true.

Number theorists, as I've said, tackle the nature of infinity and its bizarre arithmetic. Set theory, which is what we were doing when we looked at the infinite mathematical zoo, was invented by the German mathematician Georg Cantor (1845–1918). He figured out that there are actually different types of infinity. He worked on the *cardinality* of sets, which means how many members of the set there are. For instance, if I define set A as being the planets of the solar system, the cardinality of set A is eight. (For more information about why Pluto is no longer a planet, see page 132.)

Cantor looked at infinite sets too. The natural numbers are infinite, but Cantor said that they are *countably* infinite because as we count upwards from one, we are moving towards infinity, making progress. We'll never get to infinity, but we can approach it. Cantor defined the set of natural numbers as having a cardinality of aleph-zero, or $\aleph_0$ (aleph being the first letter of the Hebrew alphabet). Any other set of numbers where you can make progress also has cardinality $\aleph_0$. So if I include the negative integers with the natural numbers, I can still make progress counting through them, so the set of integers also has cardinality of $\aleph_0$.

If my set were all the rational numbers from zero to one, I could start on zero and try to work through all the fractions towards one. If I consider all the possible denominators for these fractions, I get the natural numbers again. The numerators would also be various parts of the natural numbers, so even the rational numbers from zero to one have a cardinality of $\aleph_0$. This can be extended to show that the set of all the rational numbers has cardinality $\aleph_0$.

Going back to Galileo's paradox, we can see that the set of natural numbers and the set of perfect squares both have cardinality $\aleph_0$ and hence are, in fact, the same size. Paradox no more – thanks, Cantor!

Essentially, sets with cardinality $\aleph_0$ can be methodically listed, even if that list is infinitely long. Cantor was able to think of sets which cannot be methodically listed when he considered the irrational numbers. His *diagonal argument* showed that if you write down all the irrational numbers as decimals, you can always make a new irrational number out of the ones you've written down. When you add this to the set, you can make a new irrational number from the new set. This loop means that you can never list all the irrational numbers methodically, as you keep finding ones that have been left out. Cantor said that sets like this were *uncountably* infinite and said their cardinality was $\aleph_1$.

Cantor, and many subsequent mathematicians, spent a lot of time trying to work out the relationship between $\aleph_0$ and $\aleph_1$. Cantor proposed the *continuum hypothesis*, which states that there is no set with a cardinality that is between $\aleph_0$ and $\aleph_1$ – there is nothing between countable and uncountable sets. It has since been shown that the continuum hypothesis cannot be proved, or disproved, using set theory.

What can be proved is that Cantor took a concept (infinity) that had only been considered seriously by philosophers and theologians up to that point and kick-started a new way of thinking about the very foundations of mathematics. However, the disagreements and arguments his ideas provoked caused Cantor great distress and provoked bouts of depression that plagued him for the second half of his life. We can only hope that the continuum hypothesis' inclusion as a Hilbert problem (see page 16) gave him some awareness of the greatness he had achieved. Certainly the idea that even infinities have differences is awe-inspiring stuff.

Chapter 3

ARITHMETIC

I'm going to work on the principle that you know how to count. I've never met an adult who could not count. It is the first part of mathematics that we learn, often before we go to school. Many small children can even parrot off the numbers from one to ten by rote before they have any understanding of what numbers are.

One way of looking at mathematics would be to say that it is based on understanding certain principles which can then be used to achieve certain results. Understanding and processes. However, many of us never quite get the understanding part (or it may not be offered to us in the first place) and we are left with only the process to learn. The problem with this is that, like any skill, it gets worse with neglect. Understanding also fades, but not in the same way. What I love about mathematics is that I, an unremarkable human who lives on a small island in the northern hemisphere, am at the apex of a pyramid of understanding that goes back through thousands of years, people and cultures. There are many people whose maths pyramid is far taller than mine, but I have chosen to spend my career helping other people build up their pyramids. And I know from experience that it doesn't matter how good you are at memorizing facts, algorithms and processes. Without the understanding as a foundation, at some point your pyramid is going to fall over.

Before we look at paper methods of arithmetic, I'd like to take a brief look at the dual nature of the symbols + and –. These were first introduced to the Western world in Germany from the late 1400s onwards. Johannes Widmann (*c.* 1460–98) wrote a book called, in English, *Neat and Nimble Calculation in All Trades* in 1489 which is the earliest printed use of these symbols. From the beginning, the symbols had two meanings each, which some people struggle to differentiate.

Each symbol can be either an *operation*, to add or subtract, or a *sign* to denote positive or negative. They are simultaneously an instruction and a description, a verb and a noun. +3 can mean 'add three' or 'positive three' – how do you know which is meant?

It's fairly common in mathematics education to introduce the concept of a number line – an imaginary line that helps you to perform mental arithmetic and to understand the concepts of 'greater than' and 'less than'. I often ask my students whether they see their number line as horizontal or vertical and which direction the numbers go in. I am sure that there could be some very interesting research here! For the sake of my analogy, our number line will be vertical like a thermometer.

Here we can see the use of + and – in their descriptive form, telling us whether the number is positive or negative. We don't usually include the descriptive + on positive numbers, but I've put them on here to highlight the positive part of the number line. Zero, we can see, is exactly in the middle and so is neither positive or negative.

Now, imagine you are the captain of a mathematical hot-air balloon. You have two ways of changing the height of the balloon – changing the amount of heat in the balloon and changing the amount of ballast in the balloon. We'll treat the heat as positive as it makes the balloon go upwards. You can change the amount of heat in the balloon in two ways. You can add more by using the burner, or take some away by opening a vent at the top of the balloon, allowing hot air to escape. We'll treat the ballast as negative as it makes the balloon go downwards. You can change the amount of ballast in the balloon by throwing some over the side or by having your friend with a drone deliver some more to your basket. We can represent each of these four ideas with a mathematical operation:

Action	Effect		Balloon goes. . .
Use Burner	Add +	Heat +	↑
Open Vent	Subtract -	Heat +	↓
Add Ballast	Add +	Ballast -	↓
Drop Ballast	Subtract -	Ballast -	↑

Hindu-Arabic Numerals

Our way of writing numbers is called the *Hindu-Arabic* system as it combines several breakthroughs from both these cultures. An Indian astronomer called Aryabhata (475–550) was among the first to use a place-value system from about 500 CE, specifying a decimal system where each column was worth ten times the previous. Another Indian astronomer, Brahmagupta (598–670), embellished the system by using nine symbols for the numbers and a dot to represent an empty column, which went on to evolve into our symbol for zero: 0.

The efficiency of calculation that the new system allowed made it popular and it spread across the world. By the ninth century it reached an Arabic mathematician called Muhammad al-Khwarizmi (*c.* 780 to *c.* 850) – from whose name we get the word 'algorithm' – who wrote a treatise on it. This was subsequently translated into Latin, which gave the Western world access to these numbers for the first time.

Sadly, the system didn't gain much traction in Europe. Leonardo of Pisa (*c.* 1175 to *c.* 1240), aka Fibonacci, who was educated in the Arabic world, used it in his book *Liber abaci* in 1202. The book was influential in persuading shopkeepers and mathematicians away from using the abacus for calculation and towards the awesome potential of the Hindu-Arabic system. However, it too was written in Latin, which excluded many people from understanding it. In 1522, Adam Ries (1492–1559) wrote a book in his native German explaining how to use these numerals, which finally enabled literate but not classically educated folk to exploit the system.

The last row of the table is one that many people accept (or have learnt by rote) but don't really understand why – hopefully the balloon analogy is some help!

We have now sorted out how to make our balloon go up and down, what mathematicians call an operation. If we want to calculate our altitude, our position on the number line, we need to do a calculation, which combines our current place on the number line with an operation. The first number in the calculation tells us our current altitude, and the rest of the calculation tells us what action to take. For example, we could translate –4 + 3 as:

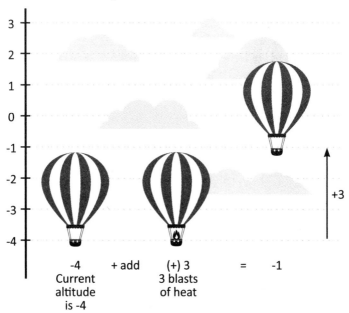

-4	+ add	(+) 3		=	-1
Current altitude is -4		3 blasts of heat			

Clearly, this means the balloon will go up three places on the number line, from –4 to –1.[†] Therefore:–4 + 3 = –1

[†] I'm fully aware that a real hot-air balloon would carry on going up, which is why I specified that this balloon is mathematical, rather than physical.

A slightly trickier example, with lots of negatives in it, would be –1 – –6, which we can translate as:

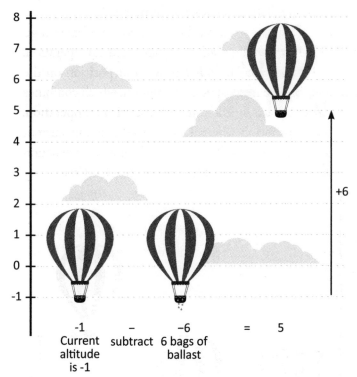

Dropping six bags of ballast over the side is going to make the balloon go up six, so: –1 – –6 = 5

Now that we know when your balloon will go up and when it will go down, we can look at more complicated arithmetic and the rest of the four operations.

Chapter 4

ADDITION AND MULTIPLICATION

When it comes to doing addition with larger numbers on paper, the methods we use all rely on the information encoded in the number by *place value*. We know that the number represented by the digits 1234 is one thousand, two hundred and thirty-four. This is because each position in the number has a corresponding value. From the right, these are ones (usually called units), tens, hundreds, thousands, tens of thousands, etc., getting ten times bigger every step to the left. So the number 1234 is four units (4), three tens (30), two hundreds (200) and one thousand (1000). I can write 1234 as:

$$1234 = 4 + 30 + 200 + 1000$$

This is called *expanded form* by maths teachers and it's really helpful for understanding how sums work. Imagine the sum 1234 + 5678. If I write each number in expanded form thus:

$$1234 = 4 + 30 + 200 + 1000$$
$$5678 = 8 + 70 + 600 + 5000$$

I can then add each matching value together easily:

1234 + 5678 :   4 + 8 = 12 (units)
                            30 + 70 = 100 (tens)
                            200 + 600 = 800 (hundreds)
                            1000 + 5000 = 6000 (thousands)

From here I can see that 1234 + 5678 = 12 + 100 + 800 + 6000 = 6912.

The way we were taught at school is merely a shorthand of this process. We set up the sum with the columns matching and add through, right to left:

	1	2	3	4
+	5	6	7	8

The first calculation is 4 + 8 = 12. We can't write 12 in the one-digit answer box but 12 = 10 + 2, so we leave the 2 in that box and carry the 10 to the next calculation:

			1	
	1	2	3	4
+	5	6	7	8
				2

Technically, the next column addition is 10 + 30 + 70 = 110, but as we are working in the tens column we can just look at how many tens we have: 1 + 3 + 7 = 11 tens altogether. So again we have too many digits to fit in. 11 = 10 + 1, so we write a 1 in the tens column and carry 1 into the hundreds column:

		1	1	
	1	2	3	4
+	5	6	7	8
			1	2

$100 + 200 + 600 = 900$:

		1	1	
	1	2	3	4
+	5	6	7	8
		9	1	2

And finally, $1000 + 5000 = 6000$:

		1	1	
	1	2	3	4
+	5	6	7	8
	6	9	1	2

Multiplication is a quick way of doing repeated addition. 12×17 asks the question: 'how much is twelve lots of seventeen?' I could work out the answer by adding twelve seventeens together, or seventeen twelves, but multiplying is much faster, provided you have learned your times tables in advance.

Imagine I had a lot of counters. I could solve the 12×17 problem by putting out twelve rows of seventeen counters, and then counting them:

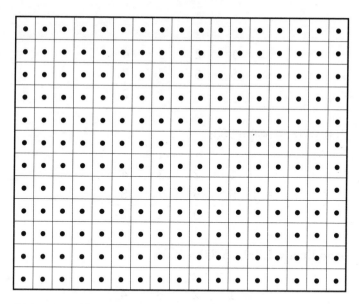

However, if I think about 12 as 10 + 2 and 17 as 10 + 7 then I can group the counters:

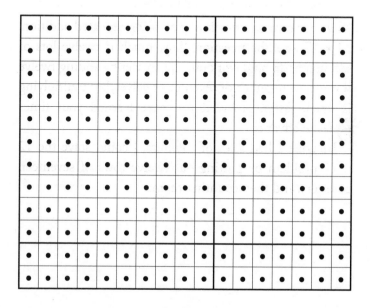

As I know my times tables, I know how many counters must be in each subdivision:

	10	7
10	$10 \times 10 = 100$	$10 \times 7 = 70$
2	$2 \times 10 = 20$	$2 \times 7 = 14$

So now I know that $12 \times 17 = 100 + 70 + 20 + 14 = 204$. This method (minus putting out 204 counters) is called the grid method. Here's a slightly more advanced version for solving 293×157:

	200		90		3	
100	$100 \times 200 =$	20000	$100 \times 90 =$	9000	$100 \times 3 =$	300
50	$50 \times 200 =$	10000	$50 \times 90 =$	4500	$50 \times 3 =$	150
7	$7 \times 200 =$	1400	$7 \times 90 =$	630	$7 \times 3 =$	21
		31400		14130		471

You might be asking how I did all the multiplications in my head when they are much larger than what we find in our times tables. Well, there's a nifty hack for that. Every time I multiply an integer by ten, I add a zero to

Decimals

It's worth noting that I can extend the idea of place value in both directions. Going to the right of the units column, the columns get ten times smaller each time, giving me tenths, hundredths, thousandths and so on. I use a decimal point to show that the right-most digit is no longer the units. This means I can use the same rules as above to add decimal numbers eg 45.3 + 27.15:

	1				
	4	5	.	3	0
+	2	7	.	1	5
	7	2	.	4	5

Notice that I put a zero on the end of 45.3 to make the columns match up, making the calculation clearer (and it's particularly important for subtraction). I can do this as 45.3 is the same as 45.30: three tenths add zero hundredths is still just three tenths. For this reason, mathematicians say 45.30 as forty-five point three zero rather than forty-five point thirty.

the end of the number. For 100×200, I know that 100 must be $1 \times 10 \times 10$ and that 200 must be $2 \times 10 \times 10$. If I put this altogether:

$$
\begin{aligned}
100 \times 200 \quad &= \underline{1} \times 10 \times 10 \times \underline{2} \times 10 \times 10 \\
&= \underline{1 \times 2} \times 10 \times 10 \times 10 \times 10 \\
&= \underline{2} \times 10 \times 10 \times 10 \times 10
\end{aligned}
$$

Remembering that every '×10' means putting a zero after the 2, I get 100 × 200 = 20000. I don't go through this entire process whenever I'm doing a grid multiplication. I just multiply the front digits and then add however many zeroes there are in the calculation to the right of it. So for 50 × 200, my thought process was 5 × 2 = 10, and then put three zeroes on. Therefore 50 × 200 = 10000. Bingo.

Back to my grid – you can see I've totalled each column. My final answer is 31400 + 14130 + 471, which I'll do an addition sum for:

		1	1		
	3	1	4	0	0
	1	4	1	3	0
+			4	7	1
	4	6	0	0	1

Final answer: 293 × 157 = 46001.

There are other methods, including long multiplication, but as long as you have a working method then stick to it. Let's move on to addition and multiplication's alter egos, subtraction and division.

Napier's Bones

John Napier (1550–1617) was a Scottish mathematician, astronomer and alchemist who invented a set of rods, known as *Napier's bones*, for doing multiplication. These contained a rod for each times table – for instance, the three-times-table rod would look like this:

If you wanted to calculate, for instance, 9 × 371, you would set the three-, seven- and one-times-table rods side by side and read across the ninth row, which would look like this:

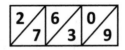

You then add together the numbers in each diagonal stripe, starting from the right. If the total is more than nine, I carry into the next stripe:

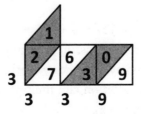

Hence 9 × 371 = 3339.

Napier was rumoured to dabble in sorcery, having a black rooster as his familiar. He would periodically command his servants to enter a room alone with the bird and stroke it, saying that this would allow the bird to sense the servant's honesty. In fact, Napier put soot on the bird's feathers. Anyone with a guilty conscience would not stroke the bird, their hands would remain sootless and they would be found guilty by the cunning Napier.

Chapter 5

SUBTRACTION AND DIVISION

Subtraction works very similarly to addition. For instance, 6543 − 5678 is:

$$6543 - 5678 : \quad 3 - 8 = -5$$
$$40 - 70 = -30$$
$$500 - 600 = -100$$
$$6000 - 5000 = 1000$$

This leaves me with $-5 + -30 + -100 + 1000 = -135 + 1000 = 865$. We can use our column method again, but whereas in addition we used carrying to cope with having too much in a column, we face the opposite problem with subtraction. If I proceed as I did above:

	6	5	4	3
−	5	6	7	8
	1	−1	−3	−5

This doesn't make a lot of sense. To find the correct answer I need to use *borrowing*, although one of my students pointed out that since the borrowed amount never gets returned, *stealing* might be a better word for it.

When I notice that 3 – 8 will give me a negative, I boost the 3 by borrowing from the next column. I cross out the 4 and reduce it by one. The 'one' I have borrowed is actually worth ten, so it increases my 3 to 13. 13 – 8 = 5:

$$
\begin{array}{ccccc}
 & & & 3 & \\
 & 6 & 5 & \cancel{4} & {}^{1}3 \\
- & 5 & 6 & 7 & 8 \\
\hline
 & & & & 5 \\
\end{array}
$$

The next column again would leave a negative result as 3 – 7 = –4. Again I borrow one of the hundreds from next door. One hundred is ten tens, boosting my 3 tens up to 13 tens so that I can proceed:

$$
\begin{array}{ccccc}
 & & 4 & {}^{1}3 & \\
 & 6 & \cancel{5} & \cancel{4} & {}^{1}3 \\
- & 5 & 6 & 7 & 8 \\
\hline
 & & & 6 & 5 \\
\end{array}
$$

Yet another borrow required for the hundreds column before I can carry on, and I can see that my thousands column will be zero:

$$
\begin{array}{ccccc}
 & 5 & {}^{1}4 & {}^{1}3 & \\
 & \cancel{6} & \cancel{5} & \cancel{4} & {}^{1}3 \\
- & 5 & 6 & 7 & 8 \\
\hline
 & 0 & 8 & 6 & 5 \\
\end{array}
$$

So we can now see that 6543 – 5678 = 865.

We saw in the previous section that addition and multiplication are closely related. The same is true of subtraction and division. The calculation 3780 ÷ 15 is asking us 'how many times does fifteen go into 3780?', i.e. 'how many times can I subtract fifteen from 3780?' Indeed, this way of thinking is the key to a method of division called *chunking*. In it, I keep subtracting multiples of the divisor until I get down to zero.

In the first place, I know that 2 × 15 = 30, so 200 × 15 must be 3000. I'll start by subtracting this from 3780:

	3	7	8	0	
−	3	0	0	0	200
		7	8	0	

This leaves 780. Thinking about fifteen, I can see that 4 × 15 = 60, so 40 × 15 = 600. I'll take this off next:

	3	7	8	0	
−	3	0	0	0	200
		7	8	0	
−		6	0	0	40
		1	8	0	

Finally, I'll take off a further twelve fifteens in two goes:

	3	7	8	0	
−	3	0	0	0	200
		7	8	0	
−		6	0	0	40
		1	8	0	
−		1	5	0	10
			3	0	
−			3	0	2
				0	

Now I know that I took away 200 + 40 + 10 + 2 = 252 lots of 15, so 3780 ÷ 15 = 252. You can see that the better you are at multiplying, the fewer steps you can chunk in.

The feared method of *long division* works along very similar principles. I set up the problem in what I call a *bus stop*:

$$15 \overline{)3\ 7\ 8\ 0}$$

I start from the left. As 15 has two digits, I look at the 3 and the 7 – how many times does 15 go into 37? Twice, giving 30, and I calculate the remainder using subtraction:

$$
\begin{array}{r}
2 \\
15 \overline{)3\ 7\ 8\ 0} \\
-\ 3\ 0 \\
\hline
7
\end{array}
$$

I now shift my attention to the 7 I have just calculated and the 8, which I'll rewrite alongside the 7. Fifteens into 78? Well, five fifteens are 75 ...

```
          2   5
  15 | 3   7   8   0
   -   3   0   ↓
             7   8
       -   7   5
             3
```

Finally, I bring the zero down alongside and consider how many fifteens are in 30:

```
          2   5   2
  15 | 3   7   8   0
   -   3   0   ↓   |
             7   8   |
       -   7   5   ↓
                 3   0
             -   3   0
                     0
```

Short division is the same as long division except that we calculate the remainders in our heads and write them in as carries. Short division is handy for converting fractions into decimals. If I want to know what $\frac{5}{8}$ is as a decimal, I can calculate $5 \div 8$:

```
  8 | 5
```

Eight goes into 5 zero times with a remainder of 5. I don't have anywhere to write this remainder until I write a decimal point and another zero. I'm allowed to do this because 5 = 5.0, and I'll write a matching decimal point above:

$$
\begin{array}{r}
0. \\
8 \overline{\smash{\big)}\, 5. \quad {}^5 0}
\end{array}
$$

Eight into 50 goes 6, remainder 2 (I can keep adding zeroes after the decimal point as needed):

$$
\begin{array}{r}
0. \quad 6 \\
8 \overline{\smash{\big)}\, 5. \quad {}^5 0 \quad {}^2 0}
\end{array}
$$

Eight into 20 goes twice, remainder 4:

$$
\begin{array}{r}
0. \quad 6 \quad 2 \\
8 \overline{\smash{\big)}\, 5. \quad {}^5 0 \quad {}^2 0 \quad {}^4 0}
\end{array}
$$

Eight into 40 goes five times exactly:

$$
\begin{array}{r}
0. \quad 6 \quad 2 \quad 5 \\
8 \overline{\smash{\big)}\, 5. \quad {}^5 0 \quad {}^2 0 \quad {}^4 0}
\end{array}
$$

So we now know that $\frac{5}{8}$ = 5 ÷ 8 = 0.625. This method works with any fraction, although you may wish to use long division for harder ones. In the next section we'll take a look at some fractions that don't work out quite so nicely.

Chapter 6

FRACTIONS AND PRIMES

We just saw how to work out a fraction as a decimal. Let's take a look at $\frac{1}{3}$ – something interesting happens:

$$
\begin{array}{r}
0.\quad 3 \quad 3 \quad 3\ldots \\
\hline
3 \,\big|\, 1.\quad {}^1 0 \quad {}^1 0 \quad {}^1 0 \ldots
\end{array}
$$

We quickly notice that a loop has formed – three into ten is three, remainder one – which will repeat for ever. Decimals that do this are called *recurring* and we use a dot to represent the digit that repeats:

$$\tfrac{1}{3} = 0.\dot{3}$$

The sevenths are even more interesting:

$$
\begin{array}{r}
0.\ 1\ 4\ 2\ 8\ 5\ 7\ 1\ 4\ 2\ 8\ 5\ 7\ 1\ 4\ 2\ 8\ 5\ 7\ldots \\
\hline
7\,\big|\,1.\ {}^1 0\ {}^3 0\ {}^2 0\ {}^6 0\ {}^4 0\ {}^5 0\ {}^1 0\ {}^3 0\ {}^2 0\ {}^6 0\ {}^4 0\ {}^5 0\ {}^1 0\ {}^3 0\ {}^2 0\ {}^6 0\ {}^4 0\ {}^5 0\ldots
\end{array}
$$

Here I get a repeating sequence of digits. I can show this by using a pair of dots at the beginning and end of the sequence:

$$\tfrac{1}{7} = 0.\dot{1}4285\dot{7}$$

What is more, every seventh uses the same sequence, just with different start and end points:

$$\frac{2}{7} = 0.\dot{2}8571\dot{4}$$
$$\frac{3}{7} = 0.\dot{4}2857\dot{1}$$
$$\frac{4}{7} = 0.\dot{5}7142\dot{8}$$
$$\frac{5}{7} = 0.\dot{7}1428\dot{5}$$
$$\frac{6}{7} = 0.\dot{8}5714\dot{2}$$

If you fancy a challenge, take a look at the nineteenths!

By looking at the denominator of a fraction I can tell whether it will recur or terminate. It all depends on whether I can take the denominator and multiply it by something to make it into a power of ten (10, 100, 1000, etc.). If I can, I can get it to sit nicely in the decimal columns when I convert it.

Before we do this, it's worth taking a look at a very important mathematical concept called *the equivalence of fractions*. It says that we can have different fractions with the same value. One way of thinking about it is with pizza. If we share a pizza, half each, we might each cut our pizza into a different number of slices, but we still have the same amount of pizza. Likewise, we all pick up the idea fairly early on at school that a half is two quarters, which is also three sixths, and so on:

$$\frac{1}{2} = \frac{2}{4} = \frac{3}{6}$$

You'll have probably been told 'whatever you do to the top, do to the bottom' by a maths teacher at some stage. What they may not have said is that this preserves equivalence.

This does give me another way to convert some fractions into decimals. For instance, $\frac{51}{250}$ is not something

I fancy doing the division for. However, if I multiply the numerator and denominator by four, I get:

$$\frac{51}{250} = \frac{51 \times 4}{250 \times 4} = \frac{204}{1000} = 0.204$$

Job done. The next thing to ponder is how can I tell whether I'll be able to multiply the denominator to get a power of ten?

To examine this, you need to understand the concept of *prime* numbers. These have fascinated mathematicians for a long time. To put it succinctly, a prime number is a natural number with exactly two factors. Eight, for instance, is divisible by one, two, four, and eight itself; four factors mean it is not prime. Five has two factors, one and five, so is prime. One has one factor – one – so is not prime. Forget that nonsense about 'a number divisible only by itself and one' for this very reason. So the first few prime numbers are 2, 3, 5, 7, 11, 13, 17, 19, 23.

One of the reasons that prime numbers are so awesome is called *the fundamental theorem of arithmetic*, which says that every natural number can be written as the product of prime numbers, but only in one way. E.g.:

$$30 = 2 \times 3 \times 5$$

There is no other combination of prime numbers multiplied together that will make thirty. Two, three and five are called the *prime factors* of thirty. For me, this makes prime numbers like mathematical DNA – every number is unique, and in numbers we don't have twins or clones to worry about! Even a huge number like 223,092,870 can only be made up one way with prime numbers (it's 2 × 3 × 5 × 7 × 11 × 13 × 17 × 19 × 23, in fact).

How does this help me with fractions? Well, I said that for a fraction to terminate I must be able to convert its denominator into a power of ten. The prime factors of ten are given by:

$$10 = 2 \times 5$$

To get the prime factors of one hundred, it helps if I recognize that:

$$100 = \underline{10} \times \underline{10}$$
$$= \underline{2 \times 5} \times \underline{2 \times 5}$$

So the prime factors of ten are two and five, and the same for one hundred (just more of them). We can see that the only prime factors of any power of ten will be two and five. Therefore, if my denominator's prime factors are some combination of twos or fives, there will be a way to multiply it to get a power of ten. My example above had a denominator of 250, and:

$$250 = 2 \times 5 \times 5 \times 5$$

So only twos and fives. I multiplied by four above, which is 2×2, to get 1000. If the denominator had been 240:

$$240 = 2 \times 2 \times 2 \times 2 \times 3 \times 5$$

This time we have a three in there, so any fraction in its simplest form that has a denominator of 240 will recur. E.g.:

$$\frac{73}{240} = 0.3041\dot{6}$$

On the other hand:

$$\frac{120}{240} = \frac{120 \div 120}{240 \div 120} = \frac{1}{2} = 0.5$$

This fraction, in its simplest form, no longer has a denominator with something other than two or five, so terminates.

Adding and Subtracting Fractions

While we are on the subject of fractions, a recap of their arithmetic is in order. To add or subtract, we need to convert the fractions so that they have the same denominator. To do this most efficiently, we look for the lowest number that both denominators are factors of – the *lowest common multiple*. For example, if I want to add five-eighths and seven-twelfths, I need to identify the lowest number that eight and twelve both go into. We quickly spot that twenty-four is on both the eight and the twelve times table:

$$\frac{5}{8} + \frac{7}{12}$$
$$= \frac{5\times3}{8\times3} + \frac{7\times2}{12\times2}$$
$$= \frac{15}{24} + \frac{14}{24}$$
$$= \frac{29}{24}$$

This is a top-heavy or *improper* fraction as the numerator is larger than the denominator. This, for some reason, is unacceptable in mathematics until you reach A Level or equivalent. I believe this is because mixed fractions are easier to understand at a glance, though improper fractions are easier to perform calculations with. To convert an improper fraction to a mixed number, I need to recognise that $\frac{24}{24}=1$. This means that:

$$\frac{29}{24} = \frac{24}{24} + \frac{5}{24} = 1\frac{5}{24}$$

Subtraction works in a similar way:

$$\frac{5}{9} - \frac{1}{4}$$ 36 is the lowest common denominator

$$= \frac{5\times4}{9\times4} - \frac{1\times9}{4\times9}$$ Use equivalence to convert to 36ths

$$= \frac{20}{36} - \frac{9}{36}$$

$$= \frac{11}{36}$$

Multiplying and Dividing Fractions

Multiplying is straightforward – I multiply the numerators and I multiply the denominators. For example:

$$\frac{3}{5} \times \frac{1}{2} = \frac{3 \times 1}{5 \times 2} = \frac{3}{10}$$

It's worth noting that when you multiply by a fraction the total gets smaller. Also, I chose a half here to highlight something that helps us divide fractions. We see above that multiplying by a half is the same as dividing by two, and likewise multiplying by a third would be the same as dividing by three. This relationship is called a *reciprocal*. Two and a half are reciprocals of each other, and it is clear if I write two as a fraction exactly how it works:

$$\frac{1}{2} \text{ is the reciprocal of } \frac{2}{1}$$

This is really handy, as it means that dividing by a number is the same as multiplying by its reciprocal:

$$5 \div 3 = 5 \times \frac{1}{3}$$

I can use this to divide fractions:

$$\frac{2}{3} \div \frac{5}{8}$$
$$= \frac{2}{3} \times \frac{8}{5}$$
$$= \frac{2 \times 8}{3 \times 5}$$
$$= \frac{16}{15}$$
$$= 1\frac{1}{15}$$

Fifteen's prime factors are three and five and so $1\frac{1}{15}$ will be recurring as a decimal.

Finding Prime Numbers

One of the reasons prime numbers have received a lot of attention from mathematicians, apart from the fundamental theorem of arithmetic, is that no one has discovered a pattern or formula for the prime numbers yet. Many have tried. For instance, the French priest Marin Mersenne (1588–1648) calculated a sequence of numbers using this formula:

$$M_n = 2^n - 1$$

You find the first number by setting n as one, the second by setting n as two, and so on. This gives you:

1, 3, 7, 15, 31, 63, 127, 255, 511, 1023, 2047 . . .

Mersenne noticed that some of the numbers given by the formula are prime numbers, such as 3, 7, 31 and 127, which are the second, third, fifth and seventh numbers in the sequence. Two, three, five and seven are prime numbers themselves, so it seems that if you use a prime number for n, you get a prime number from the formula. But the next prime number after seven is eleven, and the formula gives $M_{11} = 2047$, which is not a prime as $2047 = 23 \times 89$.

It is difficult to identify large prime numbers by hand. For instance, M_{107} is a 33-digit number, so it is very time-consuming to check whether anything divides into it and therefore is a factor.

Enter the digital age with computers that can calculate flawlessly and tirelessly. In the 1950s early computers were finding Mersenne primes, as they are known, with hundreds of digits. In 1999 the first million-digit Mersenne prime was discovered. The current record is over 22 million digits for $M_{74,207,281}$.

Looong Multiplication

At a lecture in 1903, the American mathematician Frank Nelson Cole (1861–1926) wrote the following fact about M_{67}, which was believed to be prime:

$$147,573,952,589,676,412,927 =$$
$$193,707,721 \times 761,838,257,287$$

He then proceeded to multiply this out, by hand, to prove the result. This took him an hour and was conducted in total silence. At the end of his 'lecture' Cole returned wordlessly to his seat while receiving a standing ovation from his peers.

Why bother? Well, mathematicians will always investigate anything for the sheer love of their subject. But prime numbers are also the backbone of modern-day encryption methods. If I want to transmit a number, such as my credit card details, across the internet, it is easy for people who know what they are doing to intercept this number and spend my money.

To avoid this, the internet uses a method of encryption where a *public key* is used to change the number being transmitted. This key is a combination of very large, apparently random numbers that are, in fact, created from very large prime numbers. Only the intended recipient, who has the *private key*, can reverse the process in any sort of sensible time frame.

The 'https' at the beginning of web addresses means that the website uses Hypertext Transfer Protocol with Transport Layer Security (the 's' on the end) to encrypt the information going to and from your computer. So you can happily order things online thanks to some very clever mathematics.

Chapter 7

BINARY

I mentioned computers in the first chapter and how we can outsource the fiddly activity of arithmetic to them. They are, however, the product of a lot of human ingenuity – in particular, making computers *digital*, which invariably means using electronics to make machines that can count in *binary*. This is a number system where the column values are powers (see page 15) of two (1, 2, 4, 8, 16, etc.), as opposed to the powers of ten (1, 10, 100, 1000, etc.) in the decimal system we humans use for counting.

Decimal	Binary			
	Column value: 8	4	2	1
1				1
2			1	0
3			1	1
4		1	0	0
5		1	0	1
6		1	1	0
7		1	1	1
8	1	0	0	0

The reason for this is that it is much easier, from an electronics point of view, to see voltages as being either zero or not-zero, with not-zero counting as one. If we tried to set up an analogue decimal system with zero volts for zero, one volt for one, and so on, we'd get into trouble as the resistance of the components in the computer would change as it got hotter as well as voltages dropping depending on the length of wire between components.

You'd think that, what with binary being the counting system of modern computers, it would be a fairly modern invention. It's not. Various cultures around the world used binary in all sorts of different situations. The *I Ching* (or *Book of Changes*), which has been in use for telling fortunes in China since the eighth century BCE, uses symbols called trigrams and hexagrams that are built out of binary *yin* and *yang* symbols. The great German mathematician Gottfried Leibniz (1646–1716) was fascinated by the I Ching and devised the modern binary system as a result in the late 1600s.

From here, the British logician George Boole (1815–64), in his book *The Laws of Thought*, worked out a system of logic that used binary numbers. Now known as Boolean logic, the American mathematician Claude Shannon (1916–2001) became the first person to use it in an electronic circuit in 1937 and showed that it can be used to perform arithmetic and logical operations. Shannon used switches to represent the binary information, with a switch being off representing zero and on representing one.

During the Second World War he met the British mathematical hero Alan Turing (1912–54) to discuss the use of computers in breaking Nazi ciphers. They discovered that their work complemented each other's. Shannon's 1948 paper entitled 'A Mathematical Theory of Communication' effectively makes him the father of all modern digital computers.

If you want to perform arithmetic like a computer, you'll be pleased to know that the rules remain exactly the same, as long as you remember that 1 + 1 = 10 in binary. For instance, 101 + 110 would look like this:

		1	0	1
+		1	1	0
	1	0	1	1

1010 – 111 would look like this, remembering that 10 – 1 = 1:

		1	^{1}0	
	1̶	1̶0	1̶0	^{1}0
–		1	1	1
	0	0	1	1

101 × 110 would look like this (note that you don't really need your decimal times table knowledge at all!):

×	100		10		0	
100	100 × 100 =	10000	100 × 10 =	1000	100 × 0 =	0
0	0 × 100 =	0	0 × 10 =	0	0 × 0 =	0
1	1 × 100 =	100	1 × 10 =	10	1 × 0 =	0
		10100		1010		0

And 10100 + 1010 = 11110.

For division, 1010 ÷ 100 could look like this:

```
                1  0.  1
       100 | 1  0  1  0.  0
         -   1  0  0  ↓   ↓
                   1  0   0
                -  1  0   0
                            0
```

Notice that we can have fractions in binary too and the columns right of the decimal point would be halves, quarters, eighths, etc. So the 0.1 here represents a half.

I leave it as an exercise for the reader to convert all the calculations here into decimal to check them!

The limiting factor on how quickly a computer can perform calculations is the number of switches (or *transistors*) that can be packed into a microchip, as well as dealing with the heat they produce. At the time of writing, it was possible to put over seven billion transistors in a small, commercially available computer chip. It was noted in 1965 by the American entrepreneur Gordon Moore (b. 1929), co-founder of Intel Corporation, that the technological advancement of microchips allowed the number of transistors on them to double every two years, a process dubbed Moore's law.

Moore's law has stalled in the last five years as we reach the physical limit of what we can do. Transistors produced now are only nanometres in size, meaning that you could fit millions of them in the full stop at the end of this sentence. So where do we go from here?

One possibility is to develop *quantum* computers that rely on the strange effects of quantum mechanics, which would theoretically allow them to work much faster than traditional digital computers.

It's worth noting that computers are only as effective as the humans who create them and sometimes the difference between binary and decimal can mean life or death. On 2 August 1990, Iraq began the invasion of Kuwait, beginning a conflict that would become known as the Gulf War. Iraq, led by Saddam Hussein, refused to leave Kuwait until Israel conceded various 'occupied' territories and by 17 August a coalition of thirty-four members, including the USA and the UK, launched Operation Desert Storm to liberate Kuwait.

Part of Iraq's arsenal was Scud ballistic missiles, developed by the Soviet Union during the Cold War. Ballistic missiles are rockets that, after using their fuel to propel them out of earth's atmosphere, fall back down onto their targets under gravity. They had a range of a few hundred kilometres, and Iraq used them to attack Israel, Saudi Arabia and other targets.

The US military sent several Patriot missile batteries to various locations. Patriot missiles are very fast and agile and, combined with high-precision radar, were theoretically able to destroy Scuds in the air.

To hit a moving Scud missile, the Patriot system uses radar information to predict the Scud's speed and direction, so that it can compute a trajectory and work out where to send the Patriot missile. This is not too difficult as by the time the Patriot system detects the Scud, it is simply falling under the influence of gravity.

Precision timekeeping is critical for this to work properly.

The Patriot software used a clock that kept track of time in tenths of a second. This meant that ten times per second the system was using the radar information to track where the Scud was going and update the predicted trajectory.

We saw earlier that computers use the binary system. To work out a tenth in binary, I need to do $1 \div 1010$. I'll spare you the binary long division and tell you that a tenth in binary is a recurring decimal:

one-tenth = 0.00011001100110011001100110011 …

I can write this as $0.0\dot{0}01\dot{1}$ using the notation we saw on page 40.

The computer that the Patriot system used could handle numbers with up to twenty-four digits. This seems pretty precise but, as the binary tenth goes on for ever, it meant that they had to cut off the tail of the number. Going back into the decimal system, this meant that rather than measure a tenth (or 0.1) of a second, the actual time elapsed was 0.09999991 of a second, an error of 0.00000009. This is minuscule, but as time wore on the error grew and grew as the clock was fast by this much every tenth of a second.

On 25 February 1991 the computer running the Patriot system in Saudi Arabia had been running for about a hundred hours. There are 3.6 million tenths of a second in a hundred hours, giving an overall error of about a third of a second.

Again, this does not seem to be a huge error. However, a Scud, having fallen from space, has a speed of about 1.5 kilometres *per second*, meaning that a Scud can travel 500 metres during the error time.

This meant that a Patriot missile would be launched at the wrong place or even could not be launched at all, as the radar would not find the Scud where it expected it to be and assume that a false detection had occurred.

The latter is what happened in Saudi Arabia on that day. A Scud missile hit a US barracks, killing twenty-eight soldiers and injuring many more. All due to a rounding error.

For this reason, it is important to be good at rounding, which we'll explore in the next chapter.

Chapter 8

ACCURACY

The devil is in the detail, and in this day and age we have so much detail at our disposal. Mathematicians and scientists often deal with data or measurements that have a high degree of *precision*.

For instance, the mathematical constant π is (at the time of writing) known to over 22 trillion digits thanks to the German particle physicist Peter Trueb. However, if you were trying to work out how much compost you would need for a circular flowerbed, that degree of precision would be entirely unnecessary. Even the calculator on my smartphone has twelve decimal places of π. This degree of precision would give me the area with far more decimal places than I can practically use.

Please note that being precise is not the same as being *accurate*, although we tend to use the two words interchangeably in everyday English. If I said I drank 2.7345 units of alcohol last night, this is very precise, but if I actually drank 3.2 units of alcohol, it isn't accurate at all.

In mathematics, when we want to reduce the degree of precision, while still being as accurate as possible, we use a process called *rounding*. There are several methods of rounding.

The first way that we learn at school is where we round *to the nearest*. We then specify the nearest *what*. For instance, 43 is 40 rounded to the nearest ten. 2893 is 2900 to the nearest hundred, or 3000 to the nearest thousand. The larger the number, the more we tend to round it. House prices are usually rounded to the nearest thousand, or even five or ten thousand. 'Trident nuclear submarines to cost £31 billion' has more impact than 'Trident nuclear submarines to cost £31,264,358,769.73'.

The concept behind rounding is to give lower precision while still being as accurate as possible. If I am going to round 57 to the nearest ten, I have to choose between giving my answer as 50 or 60. We choose 60, though, because it is closer to 57 than 50 is, so 60 is the more accurate choice.

If I'm rounding 250 to the nearest hundred I run into a small problem. My options are 200 and 300, but 250 is exactly halfway between them. So which one do I choose? The convention is that we always round up in these situations, so 300 would be the correct answer.

Why is the convention like this? One argument is that it means less checking when I round. If I'm rounding 250 to the nearest hundred, the moment I check the 5 digit I know to round up. If we rounded down, I'd need to check the next digit too, to see whether it was zero or not. So it's slightly more efficient this way.

Once you've mastered rounding to the nearest, the usual progression is then to learn how to round to a certain number of decimal places. This has the same thought process as before. If I want to round 1.234 to two decimal places, I still want to make sure I choose the most accurate answer. 1.234 is between 1.23 and 1.24 (some people find it easier to think of 1.234 being between 1.230 and 1.240). It is closer to 1.23.

The shortcut here is to look at the third decimal place. If it is 0, 1, 2, 3 or 4 we round down, which means leaving the second decimal place unaltered. If the third decimal place is 5, 6, 7, 8 or 9 we round up, increasing the second decimal place by one.

If you are rounding a 9 digit, you'll have to change the preceding digit too. For instance, if I round 1.96 to one decimal place, the 6 tells me to round the 9 up to 10, which means writing down a zero and carrying the one, leaving me with 2.0. Checking, I see that 1.96 is between 1.9 and 2.0 and is closer to 2.0.

Thus far we have been rounding to the nearest and to a certain number of decimal places. However, if I have a calculation to perform that contains numbers of different sizes, I don't yet have a rounding rule that provides the same degree of precision to all the numbers I'm using. For instance, if I am working out 5234×0.726, I can't round them both to the nearest hundred, as 0.726 would be zero, and I can't round 5234 to any decimal places because it doesn't have any.

So there is one last rule, called *significant figures*, that provides for this.

We count significant figures in a number from the left, ignoring any of the leading zeroes we might find in decimals. So the first significant figure of 5234 is the 5 and the first significant figure of 0.726 is the 7.

If I round these numbers 'to one significant figure', this means that I want my answer to have one non-zero digit in it (discounting any leading zeroes).

In the case of 5234, it means I am considering the 5 in the thousands column. So when I round 5234 to one significant figure I am rounding it to the nearest thousand as this is where the first significant figure is. Therefore, 5234 is 5000 to one significant figure.

Likewise, the first significant figure of 0.726 is the 7 in the tenths column, letting me know to round the number to the nearest tenth. 0.726 is therefore 0.7 to one significant figure. So my calculation is now:

$$5234 \times 0.726 \approx 5000 \times 0.7 = 3500$$

You can see that rounding to one significant figure also made the calculation much more straightforward, so the process is also used in *estimation*. Estimation is useful for checking calculations. For instance, if I had done 5234 × 0.726 by hand and got the answer 379.9884, my estimate informs me that this answer is incorrect as I am expecting something in the region of 3500. Checking, I see that 5234 × 0.726 is actually 3799.884, so I made the classic mistake of putting the decimal point in the wrong place. Estimating before calculating is a really useful sort of autocorrect for doing sums!

Chapter 9

POWERS OR INDICES

Powers (or indices, as they are also known) are seldom glimpsed in everyday life, but they have a very powerful (hence the name!) effect on the size and meaning of numbers. Powers are used to help us write down very large numbers (such as how much space your hard drive has) or very small numbers (such as how much of an active ingredient there is in a vitamin tablet).

If we write down 5^3, the superscript three is a shorthand way to write that we want to multiply the five by itself three times in total. So:

$$5^3 = 5 \times 5 \times 5 = 125$$

We would say 5^3 as 'five to the power of three' or 'five cubed'. It is very commonly mistaken as meaning 5×3, which at fifteen is a much smaller number. The powers of ten are familiar to us as they are the same as the place-value columns we use in the Hindu-Arabic numeral system (see page 23):

$10^6 = 10 \times 10 \times 10 \times 10 \times 10 \times 10 = 1{,}000{,}000$
(one million)

$10^5 = 10 \times 10 \times 10 \times 10 \times 10 = 100{,}000$
(one hundred thousand)

$10^4 = 10 \times 10 \times 10 \times 10 = 10{,}000$ (ten thousand)

$10^3 = 10 \times 10 \times 10 = 1000$ (one thousand)

$10^2 = 10 \times 10 = 100$ (one hundred)

If we continue the pattern with the powers, it follows that:

$$10^1 = 10 \text{ (ten)}$$

This shows a general law of powers, which is that anything to the power of one is just itself:

$$a^1 = a$$

The powers of ten have turned out to be so useful that we also have prefixes for words to indicate that we want to use a certain power of ten. An example is 'kilo-', which stands for 10^3 or 1000. Thus, a kilometre is a thousand metres, a kilogram is a thousand grams, and a kilowatt is a thousand watts. Other prefixes for large numbers tend to go with every third power of ten:

10^3: thousands – kilo- (k)
10^6: millions – mega- (M)
10^9: billions – giga- (G)
10^{12}: trillions – tera- (T)

We use these last few to describe the power of our computers but I expect we will see peta ($P, 10^{15}$) and exa ($E, 10^{18}$) come into common usage in the not too distant future.

Really Big Numbers

The number 10^{100} was called a *googol* in 1920 by the nine-year-old nephew of the American mathematician Edward Kasner, who wanted a name for a ridiculously large number. And it is *ridiculously* large – the universe has 'only' around 10^{80} atoms in it, meaning that a googol is 100,000,000,000,000,000,000 times larger! If that weren't enough, there is also a number called a *googolplex* which is googol^{googol}. A googol has a hundred zeroes; a googolplex has a googol zeroes.

Mathematicians have noticed another couple of shortcuts that help when you are trying to multiply or divide numbers with powers. Take a look at:

$$5^3 \times 5^4 = (5 \times 5 \times 5) \times (5 \times 5 \times 5 \times 5)$$

The brackets are only there to show the fives that make up 5^3 and 5^4. We can see that the answer is seven fives multiplied together, or 5^7. We can get to this quickly, without writing out a lot of fives, by noticing that the seven comes from adding the powers together:

$$5^3 \times 5^4 = 5^{3+4} = 5^7$$

To generalize, if you are multiplying powers of the same number together, the rule is:

$$a^n \times a^m = a^{n+m}$$

A very similar idea works with dividing. For instance:

$$8^5 \div 8^2 = \frac{8 \times 8 \times 8 \times 8 \times 8}{8 \times 8}$$

Now, if I start cancelling out eights:

$$\frac{8 \times 8 \times 8 \times 8 \times 8}{8 \times 8} = \frac{8 \times 8 \times 8 \times \cancel{8} \times \cancel{8}}{\cancel{8} \times \cancel{8}} = \frac{8 \times 8 \times 8}{1} = 8^3$$

We notice that I end up with the difference of the powers:

$$8^5 \div 8^2 = 8^{5-2} = 8^3$$

Or, in general:

$$a^n \div a^m = a^{n-m}$$

This law helps us to understand what a power of zero means. If I do $3^4 \div 3^4$, the difference of the powers gives us:

$$3^4 \div 3^4 = 3^{4-4} = 3^0$$

But we also know that a number divided by itself is one, so 3^0 must equal one. This gives us another law, which is that anything to the power of zero is one:

$$a^0 = 1$$

Negative Powers

If I use this rule to divide a smaller power of a number by a larger one, I get a negative power:

$$7^4 \div 7^9 = 7^{4-9} = 7^{-5}$$

If I look at it as a fraction (see page 41 for a discussion on simplifying fractions):

$$7^4 \div 7^9 = \frac{7^4}{7^9} = \frac{7 \times 7 \times 7 \times 7}{7 \times 7 \times 7 \times 7 \times 7 \times 7 \times 7 \times 7 \times 7}$$

$$= \frac{\cancel{7} \times \cancel{7} \times \cancel{7} \times \cancel{7}}{7 \times 7 \times 7 \times 7 \times 7 \times \cancel{7} \times \cancel{7} \times \cancel{7} \times \cancel{7}} = \frac{1}{7 \times 7 \times 7 \times 7 \times 7} = \frac{1}{7^5}$$

This means that $7^{-5} = \frac{1}{7^5}$, or 1 divided into 7^5 parts, which is a very small number. In general:

$$a^{-n} = \frac{1}{a^n}$$

If we go back to powers of ten:

$$10^0 = 1 \text{ (one)}$$
$$10^{-1} = \tfrac{1}{10} \text{ (one-tenth)}$$
$$10^{-2} = \tfrac{1}{100} \text{ (one-hundredth)}$$
$$10^{-3} = \tfrac{1}{1000} \text{ (one-thousandth)}$$

Again, some of the more commonly used negative powers of ten have prefixes associated with them:

$$10^{-2}: \text{hundredths} - \text{centi- (c)}$$
$$10^{-3}: \text{thousandths} - \text{milli- (m)}$$
$$10^{-6}: \text{millionths} - \text{micro- } (\mu)$$
$$10^{-9}: \text{trillionths} - \text{nano- (n)}$$

If you look on the back of a bottle of vitamins, you may find certain minerals are in microgram quantities. A nanometre is about the size of a molecule, hence the word 'nanotechnology'. It is important not to get confused and think that negative powers give negative numbers – in fact they give us positive numbers that are smaller than one.

Fractional Powers

If you thought negative powers were about as tricky as we could get, then guess again – we can also have fractional powers! Take a look at this:

Surds

$\sqrt{6}$ has no whole-number answer (it is, in fact, about 2.44948974). Square roots which are not integers form an endless, non-repeating string of decimal digits and are called *surds*. This comes from the Latin word for deaf or dumb, perhaps giving the idea that these numbers cannot be properly spoken aloud. A contradictory mathematical or logical argument is called *absurd* for the same reason.

Recalling that $a^n \times a^m = a^{n+m}$,

$$6^{\frac{1}{2}} \times 6^{\frac{1}{2}} = 6^{\frac{1}{2}+\frac{1}{2}} = 6^1 = 6$$

But think about it for a second – six to the power of a half, multiplied by itself, equals six. This reminds us of the section on squares and roots (see page 15). As $6^{\frac{1}{2}}$ has been multiplied by itself to give six, it must be the square root of six:

$$6^{\frac{1}{2}} = \sqrt{6}$$

And in general:

$$a^{\frac{1}{2}} = \pm\sqrt{a}$$

The $\pm$ symbol is there to make a subtle distinction. The square root symbol, when depicted alone (i.e. without a $\pm$ symbol), only ever refers to the positive square root. Numbers also have a negative square root, due to the fact that a negative multiplied by a negative gives a positive:

$$4 \times 4 = 16$$
$$-4 \times -4 = 16$$

So 4 and −4 are both square roots of sixteen and the ± 'plus or minus' symbol is there to say that there is a negative answer too.

It follows that other fractional powers give other types of root. If you think about a number raised to the power of a third, then three of them multiplied together would give the original number, implying that the third power is the same as the cube root:

$$4^{\frac{1}{3}} = \sqrt[3]{4}$$

Notice that there is no ± here. There is only a positive cube root of four, as three negative numbers multiplied together would yield a negative result. For instance, we know that the cube root of eight is two (because $2 \times 2 \times 2 = 8$), but $-2 \times -2 \times -2 = -8$, rather than 8.

In general:

$$a^{\frac{1}{n}} = \sqrt[n]{a}$$

Powers and Where to Find Them

Powers are often used to describe quantities with compound units, which means quantities that don't have a unit of their own. For instance, we describe length in units of metres (often using some of the power-of-ten prefixes to give km, cm, mm, etc.), but we describe an area in terms of metres squared (m^2). Metres squared is a compound unit because there is no other named unit for them, although we could use something like an acre or a hectare to avoid using a power.

You can sometimes see negative indices where the

Ares and Graces

A hectare is sometimes used as a unit of area, but few people know what one actually is. Well, an *are* is an old (but still metric) square of land measuring ten metres on each side, introduced (along with the rest of the metric system) by the French after the revolution at the end of the eighteenth century. *Hecto-* is a virtually unused power-of-ten prefix for 10^2 or 100. Put them together and you get a hectare, one hundred ten-metre-by-ten-metre squares with a total area of $100 \times 10 \times 10 = 10,000$ m². It's about two and a half acres, or just over a soccer pitch.

word 'per' (meaning 'divided by') would be used. Speed is a compound unit and the dashboard of some cars will say kmh⁻¹, rather than kph, for kilometres per hour. The latter is mathematical nonsense (the letters are just a simple way to represent the words 'kilometres per hour'), while the former is kilometres multiplied by hours to the power of minus one:

$$kmh^{-1} = km \times h^{-1}$$
$$= km \times \frac{1}{h}$$
$$= km \div h$$

Multiplying by one over a number is the same as dividing by that number (see Multiplying and Dividing Fractions on page 45), so it is a mathematically concise way of giving the units of speed for the car.

Many very important equations use powers. Einstein's $E = mc^2$ paved the way for nuclear energy and Newton's $F = Gm_1m_2r^{-2}$ allowed him to describe the motion of the planets. Fermat's last theorem involved the equation of a Greek-Egyptian mathematician called Diophantus: $a^n + b^n = c^n$ (for more information see page 150).

At the beginning of the chapter I said that powers or indices give us a good way to represent numbers with many digits, and technical folk often use something called *standard form* to show them. Standard form numbers are represented as a number between one and ten multiplied by a power of ten. For instance, the speed of light in a vacuum (the c in Einstein's equation) is 299,792,458 ms^{-1} (or metres per second). Rounded to one significant figure (see page 56), this is 300,000,000. Here we use our powers of ten to help us put it into standard form:

$$300,000,000 = 3 \times 100,000,000$$
$$= 3 \times 10^8$$

This number is much easier to use in calculations and allows us to avoid making errors as we meticulously enter all the digits of the original number. Indeed, modern scientific calculators have a $\times 10^x$ button for precisely this reason.

Going the other way, the G in Newton's equation (discussed on page 130) is $0.0000000000667408 \, m^3kg^{-1}s^{-2}$, which can be converted to 6.7×10^{-11}.

So, whether you are making astronomical or microscopic calculations, or just trying to tell your megas from your gigas, indices or powers are the tools you need.

Body Mass Index

The formula for body mass index (BMI) is:

$$\text{BMI} = \frac{mass}{height^2}$$

where mass is in kilograms and height in metres. Typical healthy values are from 18.5 to 25 kgm^{-2}. For example, a person 1.7 m tall who weighed 70 kg would have a BMI of just over 24. It is intended to give a rough idea of how heavy someone is for their height and hence whether they might need to gain or lose weight to be healthier. Many people calculate their own BMI, particularly after seeing it mentioned in a magazine or online. However, they may confuse squaring their height with multiplying it by two. The latter (unless you happen to be two or more metres tall) gives you an underestimate of your BMI that gets worse the shorter you are. Bad maths could have you reaching for the cake when you ought to be swapping it for a carrot.

2

RATIO, PROPORTION AND RATES OF CHANGE

Chapter 10

PERCENTAGES

You can't move for percentages these days. Special offers, interest rates, APRs, inflation, electoral swing, body fat, tax and alcoholic drinks are but a few areas where an understanding of percentages is crucial to modern life.

A percentage is a fraction with a denominator of one hundred, from the Latin *per centum*, 'by the hundred'. The percentage symbol, %, evolved from the way Italian merchants wrote *per cento* in shorthand. Percentages have become popular for two main reasons. One is that most currencies are decimalized, so financial calculations are straightforward using percentages. The other is that percentages are easy to compare with each other.

Imagine a world without percentages. You are trying to choose between two savings accounts, one offering an interest rate of $\frac{117}{997}$ and the other $\frac{59}{500}$. Which do you choose? Fractions are hard to compare beyond the simplest. However, if I say the two rates are 11.74% and 11.8%, you may now easily compare.

As percentages are essentially fractions, their arithmetic follows the same rules. However, we almost exclusively use multiplication when dealing with percentages.

Finding a Percentage of an Amount

My joy at reaching pay day and receiving my pay slip is often somewhat tempered when I realize that some of it will end up as tax. If I earn £25,000 per year and have to pay 23% of it in tax, how much will the tax man get? Well, they get £23 out of every £100, so I could divide my salary by 100 to see how many lots of £100 I get paid and then multiply that by 23:

$$25000 \div 100 \times 23 = £5750$$

It doesn't matter whether I divide by 100 first or multiply by 23 first, so I could also write this as:

$$25000 \times 23 \div 100 = £5750$$

This is the same as:

$$25000 \times \frac{23}{100} = £5750$$

So you can see that to find a percentage of an amount, I just multiply by that percentage. You could, of course, express the percentage as a decimal too:

$$25000 \times 0.23 = £5750$$

Percentage Increase and Decrease

Happy day – I have received a 5% pay rise! To calculate my new annual salary, I could use the method above to work out 5% of my salary and add it on. However, I can see that my new salary is 105% of my original salary. So to work out my new salary I do:

$$25000 \times \frac{105}{100} = £26,250$$

This only takes one step, rather than two for finding 5% and adding it on. The same result can be obtained by 2500 × 1.05 if you prefer decimals to fractions.

I'd prefer not to do an example of my salary decreasing, so let's do an example of a sale instead. Your favourite clothing outlet has a one-day 'everything 15% off marked price' sale – how can you calculate the new price of a £23 T-shirt? Again, you could work out 15% of the price and subtract it from the original, but this takes two steps. If I spot that a 15% decrease leaves me with 85% of the original price, I can do:

$$23 \times 0.85 = £19.55$$

Reversing a Percentage Change

I was interested to read recently that the markup on food sold in a supermarket is often as high as 35%. How could I work out the value of my £57.24 shopping before this markup?

To do this I need to recognize that the price I paid for my shopping is 135% of the original value, so dividing by 135 to find 1% and then multiplying by 100 should give me 100% of the original value:

$$57.24 \div 135 \times 100 = £42.40$$

Again, I can manipulate this to:

$$57.24 \times 100 \div 135 = £42.40$$

Therefore:

$$57.24 \times \frac{100}{135} = £42.40$$

The percentage is upside down, but recall that multiplying by a fraction is the same as dividing by its reciprocal (see page 45):

$$57.24 \div \frac{135}{100} = £42.40$$

So to find the original value from a percentage change I divide by the percentage.

Loan Repayments

Many of us require loans to buy large, expensive items, like property and cars. For the bank to make money out of this, they charge us interest, which is usually expressed as an annual percentage rate or APR.

If you took out a loan at a fixed rate of 5%, you might expect the amount you owe to reduce in even amounts over the lifetime of the loan. But it doesn't. That's because each payment you make is split into two parts. One part is repaying the interest that has accrued that month and the rest is used to reduce the outstanding amount.

As an example, imagine I took out a loan of £20,000 at a rate of 12% per year. At the end of the first month I would owe the original £20k plus 1% interest (as 12% per year would be 1% per month). This is, using the percentage increase idea above:

$$£20,000 \times 1.01 = £20,200$$

If I paid £500 per month, £200 of it would be paying off interest and the remaining £300 would reduce what I owed. So at the end of the first month I owe £20,200 − £500 = £19,700.

At the end of the second month the interest would be calculated:

$$£19,700 \times 1.01 = £19,897$$

My £500 payment is now £197 of interest and £303 reducing my balance, and my new balance is £19,397. I've reduced the amount I owe by more in the second month and this trend will increase until I pay off the loan.

This effect is particularly noticeable when you have a mortgage. The first few yearly statements can be pretty depressing when you see how little of the original loan you have paid off, but it does improve over time!

The formula that banks use for calculating loan repayments so that you have paid everything off at the end of the term is fairly scary:

$$\text{monthly repayment} = \frac{\text{monthly interest rate} \times \text{amount borrowed}}{1 - (1 + \text{monthly interest rate})^{-\text{number of months}}}$$

If I took out a mortgage of £150,000 over twenty-five years ($25 \times 12 = 300$ months) at a fixed rate of 5% (5% ÷ 12 = 0.42% per month), I would get:

$$\text{monthly repayment} = \frac{\frac{0.42}{100} \times 150000}{1 - (1 + \frac{0.42}{100})^{-300}}$$

If you are unsure of the $^{-300}$, take a look on page 61. When I feed this into an obliging calculator, I get:

$$\text{monthly repayment} = \frac{630}{0.7155965} = £880.38$$

In total, I end up repaying £880.38 × 300 = £264,114, over £100k more than I borrowed.

Bernouilli, Euler and *e*

Various mathematicians have had an interest in interest over the years. It was the Swiss mathematician Jacob Bernouilli (1654–1705) who noticed that it is not only the rate of interest but how often it is calculated that affects the repayments and the total amount you end up paying.

To show this, imagine if the bank in the example on the previous page charged interest daily. My daily interest rate would be 5% ÷ 365 = 0.0137% per day. This would make my daily repayment rate £28.80, which corresponds to paying £876 per month. So only slightly less per month, but better than a poke in the eye! Conversely, if the bank charged interest yearly, my monthly repayment would be £886.91.

As you can see, it is in your favour if the interest is calculated as often as possible. This is also true with investing. Bernouilli worked out that if the interest was calculated continuously, really you would be solving this problem:

$$\lim_{n \to \infty} (1+\tfrac{1}{n})^n$$

This looks intimidating, but introduces a couple of important mathematical concepts. 'Lim' is short for 'limit' and '$n \to \infty$' means 'as n tends to infinity'. Remember: infinity is not a number and we can't use it in calculations and formulae. So this is the mathematician's way of wondering: what happens to the value of $(1+\tfrac{1}{n})^n$ when n gets really, really big? The fraction part gets smaller and smaller as n gets bigger, reducing the increase it gives, but then increasing the power means we multiply that value by itself more times. This is exactly the trade-off we see with the interest payments – paying smaller amounts more frequently reduces the total we pay.

Bernoulli was looking at interest accruing in a bank account and discovered that the most interest you would receive in a year is your initial investment multiplied by 2.7 to the power of the interest rate. If I invested £100 at a rate of 5% per annum and was paid interest continually over the year, I would get:

$$100 \times 2.7^{0.05} = £105.09$$

This is not a massive increase on the £105 I would get if the interest were just paid at the end of the year in one instalment, but every little helps. As time went on and people were able to calculate with more precision, the 2.7 was improved with more decimal places. In 1978, the American Steve Wozniak (b. 1950), co-founder of Apple Inc., used an early Apple computer to calculate it to over 100,000 digits. We know the number, like π, goes on for ever without repeating:

2.71828182845904523536028747135266249 . . .

This value crops up in all sorts of areas of mathematics that originally seem unrelated. Leonhard Euler (1707–83), a highly eminent Swiss mathematician, had been using the letter e to represent it in his work on mechanics (how things move rather than how to fix machinery) and thus it became known as Euler's number. He also famously used it in Euler's identity:

$$e^{i\pi} + 1 = 0$$

One and zero speak for themselves, being the basis of counting and all other numbers. π represents geometry, the number that describes the essence of a circle. i is the square root of minus one, a non-existent imaginary number that nonetheless allows us to solve all sorts of otherwise insoluble equations. This pithy statement contains five of the most important numbers and is considered by many to be the most elegant – some would say beautiful – equation in the whole of mathematics.

Chapter 11

COMPOUND MEASURES

There is a technique called *dimensional analysis* that scientists, engineers and mathematicians can use to check whether their latest formula makes sense in terms of the units used. Here is the formula for the area of a square:

area of a square = length × length

Area, we know, is measured in units of metres squared, m^2. If I look at the units of the other side of the equation, the length × length part, the units of both are metres, and when multiplied together, metres times metres also gives metres squared. The units of both sides of the formula match. We have analysed the dimensions of the formula. This doesn't guarantee that the formula is correct, but it is a helpful tool.

Think about the area of a circle:

area of a circle = radius × radius

This statement is dimensionally correct for exactly the same reasons as the previous example. However, we know that we need our old friend pi in there to give us the correct answer:

$$\text{area of a circle} = \pi r^2$$

For pi not to muck up our dimensional analysis, it must have no dimensions itself. Pi is a constant with no units – we don't measure pi in terms of metres, kilograms or anything else. So pi is a *dimensionless constant*, and much effort has been put into working out values of such things over the years.

One of the best things to come out of the French Revolution in the late 1700s was the *metric system*. The revolutionaries linked together the need for a new, coherent system of weights and measures with the ease of use of the decimal system of numbers. Since the time of the Renaissance, natural philosophers (the predecessors of modern scientists and mathematicians) had been in increasing communication and needed a more universal way of communicating quantities.

The system has been refined over the years, but the essential idea was to have a small number of fundamental base units from which all other measurements could be derived. Initially, the base units were the metre (for length), the kilogram (for mass) and the second (for time). The French revolutionaries defined a metre as one ten-millionth of the distance from the North Pole to the equator through Paris. The kilogram was originally the mass of one thousand centimetres cubed of water at just above freezing. In 1899 it was redefined as being the mass of the International Prototype Kilogram, a very carefully manufactured cylinder of incredibly hard-wearing and corrosion-resistant platinum and iridium alloy. This definition is due to be updated in the next few years to something more conceptual involving fundamental physical constants. The second was a unit of time that had been in long use, its sixtieth-of-an-hour nature being a hangover from the number systems of the ancient cultures that had first measured time. Now the definition of a second is tied to the extremely consistent frequency of vibration of an atom of a metal called caesium – the time it takes to vibrate 9,192,631,770 times, to be exact.

Other base units were added as our understanding of the universe improved and needed measurement: the kelvin, the unit of temperature; the ampere, the unit of electric current; the mole, the amount of a substance in a certain mass (sadly, nothing to do with cute burrowing mammals); the candela, the unit of light intensity or brightness. The seven base units in what is now known as the SI (Système international) allow us to describe anything we need to in the physical world.

Many of the derived or compound units we use have been given names too. For instance, a *litre* is a unit of volume and is equal to a thousandth of a metre cubed, but it is far easier, bleary-eyed on the Monday morning commute, to wrap your head around litres of petrol than thousandths of a metre cubed. Many derived units are named after scientists whose work was important in that field.

Isaac Newton (1642–1726) was the English prodigy famous for developing the concept of gravity, a force.

Newton's second law: force = mass × acceleration

Mass, a basic SI unit, is measured in kilograms. Acceleration is measured in metres per second squared. Thus the units of force are 'kilogram metres per second squared'.

However, we now call this compound measure newtons.

Blaise Pascal (1623–62) was a Frenchman who, amongst other things, did a lot of work on fluids, inventing the hydraulic press and the syringe. Much of his work focused on pressure:

$$\text{pressure} = \frac{force}{area}$$

Force, as we have just seen, has units of kilogram metres per second squared and area has units of metres squared. Thus the unit of pressure should be 'kilogram metres per second squared per metre squared'. We can cancel this down to 'kilograms per metre second squared' but I'm sure you'll agree that pascals, the unit of pressure, are much easier to say.

Both the definitions above end up with 'per second' in them somewhere. When we divide by a unit of time, we are looking at how something changes in that unit of time. This is so common in science and mathematics that we call this idea a rate of change. The most common rate of change we see is speed. You may remember learning this at school, possibly with the assistance of this triangle:

$$\text{speed} = \frac{distance}{time}$$

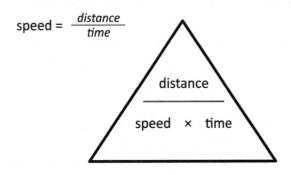

As a maths teacher, the first point I want to make is that this formula works out *average* speed. For instance, if I travel by train from York to London, a distance of 200 miles, in 2 hours, this formula says my speed must be 200 ÷ 2 = 100 miles per hour. But this is clearly my average speed as the train must have stopped to let me on, sped up, slowed down, stopped at other stations and finally terminated in London.

Speed is a rate of change. In this case, it says that in every hour I have travelled an average distance of 100 miles. My position, relative to my starting point, has changed by 100 miles.

Acceleration is the rate of change of speed – how much your speed changes by in a given amount of time. Things accelerate when we drop them – because of gravity – at 9.81 metres per second squared on Earth. 9.81 metres per second squared means that every second, the speed of the falling object increases by 9.81 metres per second. This is quite fast, which explain why pretty much everything except cats gets broken when it falls from high enough. On the Moon, the acceleration is only about 1.66 metres per second squared and so we get astronauts, wearing spacesuits that weigh more than the astronauts themselves, able to bounce around without spacehoppers.

As we experience the 9.81 metres per second squared acceleration for all of our lives, we tend to compare other accelerations to it, calling them 'G-forces'. We accelerate when we change direction as well as speed, hence the force we feel going around a bend in a car. Rollercoasters subject us to accelerations that can be up to six times Earth's gravity, or 6g. This is only for very short periods

Getting It Wrong: Mars Climate Orbiter

Several hundred years after the invention of the metric system, we still find older systems in use around the world. In the UK, we still like our beer and milk in pints (568 millilitres isn't quite so pithy), and road signs are still in miles. The United States uses a similar, but not identical, system that is also still used in some industries.

This has caused a number of embarrassing problems. In 1983 an Air Canada flight, using an American Boeing 767, ran out of fuel in mid-flight because the ground crew loaded the fuel in pounds rather than kilograms. As a kilogram is more than two pounds, this meant there was less than half the expected amount of fuel. This, coupled with a dodgy fuel gauge, meant the flight had to glide to a crash-landing on an ex-military airfield now used as a racetrack. Fortunately, no one was seriously injured and the pilots were simultaneously demoted for not checking the fuel properly and awarded for their quick thinking and outstanding airmanship.

In 1999 the Mars Climate Orbiter was due, as the name suggests, to go into orbit around Mars to study its climate after a nine-month journey from Earth. Not long after the engines were used to insert it into orbit, contact was lost. An investigation subsequently discovered that one subcontractor had used pounds, rather than the SI-derived unit newtons, to determine the thrust produced by the engines.

This was a particularly bad mistake for several reasons. First, dimensional analysis shows that pounds and newtons are not even both measures of the same thing – the newton is a measure of force whereas the pound is a measure of mass. Second, the strange performance of the engines had even been flagged up beforehand, but the complaints were ignored.

Anyway, the result was that the Orbiter entered Mars' atmosphere much too fast and broke up before it could ever be used for its intended purpose. The cost of the project was over US$600 million.

of time, though. Astronauts and fighter pilots, who may have to endure these stresses for longer, often wear a G-suit that helps to keep their blood in their brain by squeezing their body!

Decelerations also produce G-forces, and cars are intentionally designed with crumple zones. These allow the change in speed to happen over a longer period of time, reducing the overall deceleration and therefore the force we experience. Airbags do this too, giving our bodies more time to decelerate than otherwise.

Chapter 12

PROPORTION

The concept of proportionality seems obvious, but it wasn't always so. When you have a recipe for four people but you know you have eight coming for dinner, you instinctively know that you have to double all the quantities as the number of guests doubles. This is because the number of guests and the amounts in the recipe are in proportion. But what if one guest was bringing a friend, increasing the total to nine? Suddenly it's not quite so straightforward.

To a mathematician, this means that there is a relationship between the number of guests and, say, the amount of pasta I need to cook. If the original recipe calls for 300 g of pasta, I could use an 'if, then, so' or unitary method to work out how much to make for my guests:

If 4 guests need 300 g of pasta,
then 1 guest needs 300 ÷ 4 = 75 g of pasta,
so 9 guests need 75 × 9 = 675 g of pasta.

This method is called unitary because we work out the requirement for a single person, or unit, in working out the answer. Another way of looking at it would be to work out a formula for how much pasta is required. Then, no matter how many guests you had, you could conveniently work out the amount. Let's say that the number of guests was *n* and the corresponding mass of pasta was *m*. Mathematicians would start the process of working out the formula by stating this:

$$m \propto n$$

That symbol simply means 'is proportional to'. We don't have a usable formula yet, but we have stated our assumption. Now we need to add some numerical information. We saw in the unitary method that one guest requires 75 g of pasta, which I then multiplied by the number of guests to get the total I required. We now have n as the number of guests, so my formula becomes:

$$m = 75 \times n$$

The 75 in this case is called the *constant of proportionality* – it's the number that allows us to link the two variables m and n.

In the pasta example, the number of guests and the mass of pasta are in *linear* proportion. This is because the relationship could be graphed to produce a straight line that goes through the origin (the bottom-left corner of the graph, where m and n are both zero):

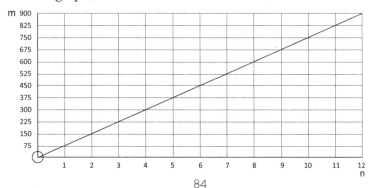

Getting It Wrong: F-22s over the Date Line

The US military F-22 Raptor is the very pinnacle of fighter aircraft. Almost impossible to detect on radar, it marries the latest developments in aerodynamics, engineering and aviation with high-performance computers that manage almost every aspect of the aircraft.

Which makes it a big problem if the computers crash.

Exactly this happened in February 2007. Twelve F-22s were travelling from a base in Hawaii to one in Japan. In doing this, they crossed the International Date Line. This is the arbitrary and somewhat wobbly line that goes between Alaska and eastern Russia, south through the Pacific Ocean, making various detours around groups of islands, past New Zealand and so on. Going from east to west, as the planes were, you lose twenty-four hours.

The US military have not released information on the exact nature of the problem, but I suspect that suddenly the planes could not compute any of the rates of change so necessary to their operation. With many systems reporting values outside their normal parameters, the planes lost their navigation and communications systems completely. Luckily, they were near their refuelling tankers at the time, which they could visually follow back to their base in Hawaii. US$148 million for a plane you can't even fly around the world!

Not everything that is in proportion produces a straight line, though. Imagine if my company manufactured and sold square tiles. The length of a tile obviously affects its area, but not in a linear way. If I double the length of a tile, I increase the area by four:

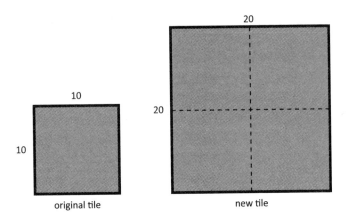

original tile   new tile

In fact, because the area of the tile is the length squared, the relationship works like this:

$$area \propto length^2$$

Both the pasta and the tiles are examples of *direct* proportion – as one value increases, so the other increases proportionally. Indirect or inverse proportion is where as one value increases, the other decreases in proportion. For instance, if I have a team of litter collectors working at a music festival, the time it takes them to clear up goes down as the number of litter collectors goes up. In this case:

$$time\ taken\ to\ clear\ up \propto \frac{1}{number\ of\ litter\ collectors}$$

If I know that it takes 10 litter collectors 50 hours to clear up, I can turn this relationship into a formula by introducing a constant of proportionality, which I need to work out. If I call it k:

$$50 = k \times \tfrac{1}{10}$$

From here I can see that $k = 500$ (as 500 times one-tenth equals 50), so my formula becomes:

$$t = 500 \times \frac{1}{n}$$

where t is the time taken and n the number of litter collectors.

Watch out for the old favourite: 'If it takes eight diggers eight hours to dig eight holes, how long does it take one digger to dig one hole?' I think the poetic part of our brain wants us to say 'one' so badly that it completely overrides the mathematical part. The correct answer is, of course, eight hours. Unitary method brings in the concept of man-hours here – how much work a person can get done in an hour. Eight diggers, eight hours, eight holes clearly implies that each digger manages a hole in eight hours. Hence, the task of digging a hole is eight man-hours' work.

Proportion, in everyday English, often means a comparison of an object's length and width. There is something intrinsically pleasing to the eye about the relative proportions of certain things. Over the centuries, many mathematicians, scientists, philosophers, architects, designers and artists have been fascinated by this phenomenon.

Euclid (fl. *c.* 300 BCE), the Daddy when it comes to geometry, was the first to describe what is now known as the *golden ratio*, which can be used to make things look proportionally lovely. He considered lines, but I think it is easier to appreciate if we look at rectangles.

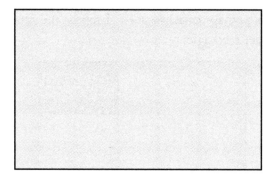

The width and height of the rectangle above conform to the golden ratio. According to many, this makes it the most pleasing rectangle to gaze upon and rectangles in this ratio were used by architects to design everything from medieval mosques to modern buildings. The Swiss architect Charles-Édouard Jeanneret (1887–1965), better known as Le Corbusier, used the ratio extensively in his work.

The key feature of this rectangle is that if I cut it into the largest square possible and a rectangle, the little rectangle has the same proportions as the original rectangle.

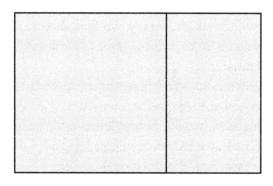

This is perhaps easier to see if I repeat the process on the smaller rectangle:

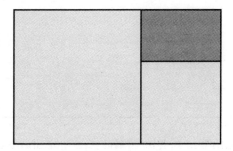

The grey rectangle and the original rectangle also have the same proportions. If this diagram reminds you of the work of Piet Mondrian (1872–1944), the Dutch artist famous for his geometrically blocked abstract paintings, it's with good reason – he wanted to draw the most pleasing shapes and so his works conformed to the golden ratio.

Why is this rectangle pleasing to the eye? No one is entirely sure but the American engineer Adrian Bejan believes that our eyes have evolved to find these proportions the easiest to absorb at a glance, allowing our brains to process the information easily. This is perhaps why, apparently unconsciously, we find shapes, pictures, buildings and even faces that conform to the golden ratio the most attractive.

The golden ratio is so important in mathematics that, like π and e, it has been given its own letter: φ (pronounced 'phi' to rhyme with 'eye'), the first letter in the name of the ancient Greek architect and sculptor Phidias (*c.* 480–430 BCE), the father of classical Greek architecture. Like π and e, φ is also an irrational number:

$$\varphi = \frac{1 + \sqrt{5}}{2} = 1.61803398\ldots$$

(For a look at why φ takes this value, see page 127.)

So our beautiful rectangles are just over one and a half times as long as they are wide. People consider faces where the height and width are in the golden ratio more attractive on average – you can even download apps that analyse photographs with facial recognition technology to determine how 'beautiful' you and your friends are!

Chapter 13

RATIO

I often think of the division symbol or *obelus*, ÷, as being a combination of the ratio symbol, :, and the fraction bar we use to draw a fraction. I have no idea whether Johan Rahn (1622–76), the Swiss mathematician who first used it, intended this, but it does serve to highlight the relationship between division, ratio and fractions. The link between division and fractions is pretty clear:

$$5 \div 7 = \frac{5}{7}$$

The ratio 5:7, however, does not have quite the same meaning as the division and the fraction. If two quantities are in the ratio 5:7, it means for every five of one, you have seven of the other.

The instructions on my couscous packet suggest using 160 millilitres of water for every 100 grams of dry couscous. Conveniently, 160 millilitres of water has a mass of 160 grams, so I could write down this ratio as:

water : couscous
160 : 100

If I made exactly this quantity of delicious Mediterranean carbohydrate, what would the fractions of water and dry couscous be? This is where ratios diverge from fractions.

The proportion of water couldn't possibly be $\frac{160}{100}$ as this is more than one whole. Equally, the dry couscous does not contribute as much as $\frac{100}{160}$ of the finished product.

Instead, we can see that my bowl of couscous must contain 160 + 100 = 260 g. This means the proportions are $\frac{160}{260}$ for the water and $\frac{100}{260}$ for the dry couscous. It is the total of the two numbers in the ratios (sometimes called the total number of parts) that gives us the denominator of the fraction.

As with fractions, we can use equivalence to modify ratios as we need to. If I reduce the couscous ratio to its lowest terms by dividing both numbers in the ratio by 20:

$$160:100 = 8:5$$

Say I am making a batch and discover, on weighing out, that I have 250 g of couscous – how much water would this require? Well, if I start with the original 160:100 ratio, I can see I have 2.5 times as much couscous, so I will need 2.5 times as much water. 2.5 × 160 = 400 g of water required.

A lot of school questions on ratio involving the sharing out of money: if Auntie Anna gives her young nieces Billy, Clara and Daisy £100 in the ratio 2:3:5, how much does each get?

The first thing to notice is that we can have ratios with more than two quantities, which we can't with fractions. The way to deal with this problem is to imagine Auntie Anna, with a pile of a hundred pound coins, handing them out. Two for Billy, three for Clara, five for Daisy. In each round of sharing she gives out £10. Clearly, she can do ten rounds of sharing in total with her £100, so each niece gets ten times their share each round, £20, £30 and £50 respectively:

B	:	C	:	D	Total Number of Parts
2		3		5	10
20		30		50	100

One of the other areas where we see ratio being used is in map scales. Simple maps may have a scale such as '1 cm represents 100 km' or suchlike. However, other maps and atlases may just use a ratio like 1:50,000. This is the ratio of the size of the thing on the map to its size in real life.

For instance, if the distance between my house and the betting shop is 5 centimetres on the map, how far is it in real life?

Well, the ratio tells me that the real distance would be 50,000 times as far. 5 cm × 50,000 = 250,000 cm. How far is that? Dividing by 100 converts centimetres to metres, giving me 2500 m or 2.5 km. Not too far.

When I get to the betting shop, I'll be confronted with more ratios, in the form of *odds*. Traditionally, even though the odds are ratios they are shown with 'to' or –. Sometimes they are even made to look like a fraction with a /. Any way you show them, the odds show you the ratio of winnings to the stake made. For instance, five-to-two (or 5–2 or 5/2) odds *against* mean that for every two pounds bet, you would win five. Your stake is also returned, so you would have £7 after a successful £2 bet. Five-to-two odds *on* means that you win £2 for a stake of £5, still leaving you with £7 if you win.

It's worth noting that these odds are not probabilities, although they will reflect the probabilities to some extent. For more information on probabilities, see page 177.

3

ALGEBRA

Chapter 14

THE BASICS

The concept behind algebra is staggeringly simple: we use letters to replace numbers that we either don't know (unknowns) or will want to be able to change (variables).

That's it. People have been doing algebra for a long time. Primary school children do it:

I'm thinking of a number. Twice my number is ten.
What's my number?

This could be a starter question in any maths lesson in the world. In solving it, no matter how you get there, you are dealing with an unknown number, an abstraction. Algebra follows exactly the same rules as normal arithmetic. Through algebra, mankind has been able to theorize about and explain the universe in unprecedented detail compared to the previous religious and philosophical interpretations. Algebra is useful in all the other areas of mathematics as well as in its own right.

How did you solve the problem above? Unless you guessed at it, I suspect you thought to yourself something like: if twice the number is ten, then the number must be half of ten. Half of ten is five!

Here, you've shown that you understand the process of doubling, or multiplying by two, and that you recognize that dividing by two is the opposite, or *inverse*, of this process. Well done you.

It was the French philosopher and mathematician René 'I think, therefore I am' Descartes (1596–1650) who first introduced the modern system of using letters in algebra. He might have looked at the problem like this:

$$2n = 10$$

This concise notation introduces the unknown number, n. 2n means 'two times n'. We could write 2 × n, but the times sign is confusingly like an x which is commonly used, so the convention is that we leave that symbol out, only putting it back in if we replace the letters with numbers. Finally, '= 10' means just that.

Linear Equations

A problem like the one above is called an equation. We can often (but not always) solve equations. One way to do this is to imagine a balance or seesaw. The equals sign represents the fulcrum or pivot. Here's the seesaw for our problem:

We can see that to match the ns and the ten, we'd need to split the ten into two equal boxes (i.e. divide by two):

Thus, n must be equal to five.

The seesaw analogy is helpful because we know from experience that to keep the seesaw level, we have to have the same weight on each side. If I change one side, I must change the other in the same way to keep things balanced, or equal.

I am thinking of a number. Three times my number plus eight is twenty. What is my number?

This one's a bit harder, but using n again, our equation would be 3n + 8 = 20. The seesaw would look like this:

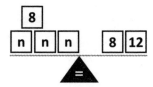

I can think of 20 as being 12 + 8:

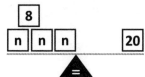

Now I can remove 8 from both sides without affecting the balance:

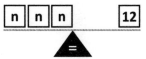

To match the ns, I have to divide the 12 into three:

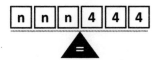

Therefore, n must equal 4. If I take each step in turn, applying it to the equation rather than the seesaw, I get:

$$3n + 8 \quad = \quad 20$$
$$(- 8 \text{ from both sides})$$
$$3n \quad = \quad 12$$
$$(\text{divide both sides by 3})$$
$$n \quad = \quad 4$$

Solving an equation is a bit like playing a game, eliminating things from the seesaw or equation until you have the unknown by itself on one side and a number on the other. We use inverse operations to do the eliminating. Here's an example with subtraction and division in it:

$$\frac{n}{5} - 2 \quad = \quad 4$$
$$(+ 2)$$
$$\frac{n}{5} \quad = \quad 6$$
$$(\times 5)$$
$$n \quad = \quad 30$$

The game gets slightly harder if you put the unknown on both sides. But only slightly:

$$5n + 7 \quad = \quad 3 + n$$
$$(- n)$$
$$4n + 7 \quad = \quad 3$$
$$(- 7)$$
$$4n \quad = \quad -4$$
$$(\div 4)$$
$$n \quad = \quad -1$$

It's always worth remembering that we can check the solutions to equations by putting the number we found back into the original equation. This is called substitution:

$$5 \times -1 + 7 \quad = \quad 3 + -1$$
$$-5 + 7 \quad = \quad 2$$
$$2 \quad = \quad 2$$

Everything checks out, so our answer is correct.

Quadratic Equations

All of the equations above are called linear equations because they contain unknowns without any powers linked to them. Any equations with a squared unknown are called quadratics and we can use different strategies to solve them.

Imagine a rectangle – it has an area of 15 m^2 but I'm not sure of the exact lengths of the sides. Can I work them out from this information?

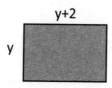

You get the area of the rectangle from multiplying the width and the length, so I could set up an equation like this, remembering that we omit the multiplication sign:

$$y(y + 2) = 15$$

So I need two numbers that multiply to make 15, where one number is two more than the other.

Three and five seem to work quite nicely here – job done. But I cannot be sure that I'll always be able to work them out in my head, so let's take a look at some algebraic ways of solving this quadratic equation.

I can't use the seesaw method here – the unknown is in two places and I can't think of a way to combine them. I can split the rectangle into a square and a smaller rectangle to help me. The area of the square is y × y = y^2. The area of the rectangle is 2 × y = 2y, as shown below:

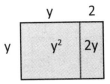

Now I can see that $y(y + 2) = y^2 + 2y$ and, because the area of the square and small rectangle is the same as the big rectangle, this must also equal 15:

$$y^2 + 2y = 15$$

So I have managed to multiply out the brackets and I can now see the telltale y^2 term – this is what makes this a quadratic equation. There are three methods of solving quadratics using algebra: completing the square, factorizing and using the quadratic formula.

Completing the Square

This method involves knowing about what happens when you multiply out or *expand* brackets like:

$$(x + 1)^2$$

Again, I can use geometry to help me because $(x + 1)^2$ means $(x + 1) × (x + 1)$: a number times itself, so the area of a square. If I split up the sides into parts of length x and length 1 I get:

If I work out the area of each section I get:

	x	1
x	x^2	1x
1	1x	1

Therefore, $(x + 1)^2 = x^2 + 1x + 1x + 1$. I can add the 1x terms together, giving me:

$$(x + 1)^2 = x^2 + 2x + 1$$

So what? Well, compare the right-hand side of this to the left-hand side of the quadratic I'm trying to solve. $x^2 + 2x + 1$ and $y^2 + 2y$ contain very similar terms. There's only a + 1 difference. So I can use this fact to say:

$$y^2 + 2y = (y + 1)^2 - 1$$

Why go to all that effort? Because I have now written the quadratic in such a way that the y is in only one place, so I should be able to solve it in the same way I did linear equations. First, let's write out the whole equation and show what we have changed it to:

$$y^2 + 2y = 15$$
$$(y + 1)^2 - 1 = 15$$

Adding one to both sides gives:

$$(y + 1)^2 = 16$$

Now I need to pause for a moment to remember how to unsquare something. Page 15 tells me that I have to square root both sides:

$$y + 1 = \pm\sqrt{16}$$

I have used the $\pm$ to remind me that there are two numbers that I can square to get sixteen, four and minus four:

$$y + 1 = \pm 4$$

Now subtract 1 from each side:

$$y = \pm 4 - 1$$

This means y could be either +4 − 1 = 3 or −4 − 1 = −5. Both are valid solutions to the quadratic equation, but only y = 3 makes sense in the context of rectangles. Now that we know that y is 3, I can work out the y + 2 side. So, as we previously did by inspection, we can see that the two sides of the rectangle must be 3 m and 5 m.

Factorizing

The factorizing method relies on the fact that if I have two numbers that when multiplied together give zero, one or the other of them must be zero:

$$\text{If } m \times n = 0$$
$$\text{then either } m = 0 \text{ or } n = 0 \text{ or both}$$

My equation is $y^2 + 2y = 15$, so the first thing I need to do to exploit this is to make both sides equal zero, by subtracting 15 from both:

$$y^2 + 2y - 15 = 0$$

Factorizing a quadratic involves writing it as two brackets multiplied together, like so: $(y + n)(y + m)$, where n and m are numbers. Again rectangles can help me:

I can see here that $(y + n)(y + m) = y^2 + my + ny + nm$. If I compare this to my equation, I can see that I need my + ny to make 2y and nm to be –15:

$$y^2 \qquad + 2y \qquad - 15$$
$$y^2 \qquad + my + ny \qquad + nm$$

So I need two numbers that add to make two and multiply to make –15. The only way for two numbers to multiply to make a negative and add to make a positive is if there is a larger positive number (to make the sum positive) and a smaller negative number (to make the product negative). In this case, it works if m = –3 and n = 5. Then m + n = –3 + 5 = 2 and mn = –3 × 5 = –15 as required.

I now know that:

$$y^2 + 2y - 15 = (y - 3)(y + 5) = 0$$

For this to be the case, either y – 3 = 0 or y + 5 = 0. These are two pretty easy examples of linear equations, which I can solve in my head: y must be either 3 or –5, just like we got with completing the square. Again, both are valid solutions but only y = 3 makes sense in terms of a rectangle.

In general, to solve a quadratic equation by factorizing we are looking for two numbers that have a sum that equals the number in front of the y term and a product equal to the constant (i.e. without any y) term. Use these in your brackets, think about each one being zero and you're done.

The Quadratic Formula

It's not always easy to complete the square and it's not always obvious how to factorize a quadratic. Fortunately for us, people have used completing the square in general terms to give a formula to solve quadratics. Brahmagupta (597–668) was an Indian mathematician who came up with one of the first, but as Descartes and his notation were still a thousand years away, his formula was in words rather than letters. Anyway, for a quadratic equation of the form $ay^2 + by + c = 0$:

$$y = \frac{-b \pm \sqrt{b^2 - 4ac}}{2a}$$

This looks alarming, but is actually fairly easy to use. Our equation was $y^2 + 2y - 15 = 0$, so we can see that a = 1, b = 2 and c = –15. Substituting these into the formula gives:

$$y = \frac{-2 \pm \sqrt{2^2 - 4 \times -15}}{2 \times 1}$$

$$y = \frac{-2 \pm \sqrt{4 - -60}}{2} = \frac{-2 \pm \sqrt{64}}{2}$$

$$y = \frac{-2 \pm 8}{2}$$

$$y = \frac{-2 + 8}{2} = 3 \text{ or } y = \frac{-2 - 8}{2} = -5$$

There's a couple of things to note here. First, the formula allows us to solve quadratics where the squared term begins with a number. It's possible to do this with completing the square and factorizing, but it makes it quite a bit more complex. Second, if the $b^2 - 4ac$ part is negative, you can't find the square root of it, which means that the quadratic equation has no solutions. Real solutions, that is. Mathematicians have introduced the idea of the square root of –1 being *i*, the imaginary number. Using this gives all sorts of interesting results with a surprising number of non-imaginary applications in fields such as electronics and quantum physics.

Vocabulary

There are a few key words that it is useful to understand as you take algebra further. Some have appeared above, but to sum up:

Key Word	Definition	Example
coefficient	A number in front of an unknown	$3x^2$: 3 is the coefficient of x^2
expression	Some algebra without an =	$3x^2 + 2x - 5$
term	Part of an expression	$2x$ is a term in the expression above
equation	Some algebra with an =. We may be able to find some solutions.	$3x^2 + 2x - 5 = 0$ is a quadratic equation
formula	An equation that we substitute values into that generally expresses a physical law	$E = mc^2$ is a formula that links energy (E), mass (m) and the speed of light (c)
identity	An equation that is true for any value of the variables in it. The triple bar ($\equiv$) can be used to show an identity.	$a(b + c) \equiv ab + bc$ This is true for any value of a, b and c.

Descartes' Letters

Descartes was not the first to use letters to represent unknowns, but in his 1637 work *La Géométrie* he was the first to set out the modern way of doing algebra. In this work, he was also able to link two fields of mathematics – geometry and algebra – together for the first time. This allowed geometric shapes and lines to be expressed as algebraic equations. Legend has it that Descartes first came up with the x-y coordinate system (also known as Cartesian coordinates, after him) when lying in bed watching a fly move about on the ceiling. He wanted to be able to describe the position of the fly accurately, so chose a corner of the ceiling as his starting point (or *origin*) and used two numbers as *coordinates* to locate the fly, much like the game Battleships.

Chapter 15

OPTIMIZATION

Many problems faced in real life are ones of optimization –
trying to make some things as large as possible (often profit,
sales, views or hits) and others as small as possible (costs,
time). If the problem can be described mathematically,
using algebra, then there's a good chance we can also use
algebraic techniques to find the ideal solution.

There is an old legend that says a knight with many
sons would, as a wedding gift to them, give as much of
his land as they could enclose with a 100-yard rope. He
stipulated that they could use one of his existing fences
for one boundary and that the area had to be rectangular.

How could one of his more mathematically gifted sons
ensure he got the most land possible? We can model the
situation like this:

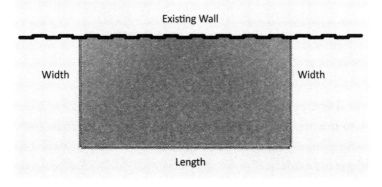

Existing Wall

Width

Width

Length

We know that the rope is 100 yards long, so width + length + width = 100. If I use w for width and l for length I get $2w + l = 100$. I can also say that the length must be $100 - 2w$, which will be useful in a minute.

I also know that I want to maximize the area. The area is the width times the length. If I call the area A:

$$A = wl \text{ but } l = 100 - 2w \text{ so}$$
$$A = w(100 - 2w)$$

I now have a formula for the area, depending on what I set the width as. At this point, I could start trying out values of w, seeing which gives me the most, but I couldn't be sure I'd got the maximum value unless I tried every possibility. One option would be to plot a graph:

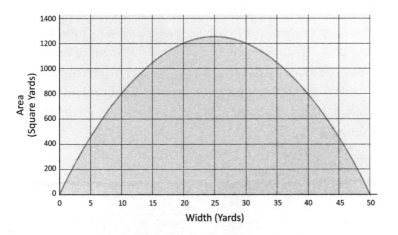

Here I can see that there is definitely a maximum value to the area – 1250 square yards when w is 25 yards. This is called a *graphical* solution, but is there a way to find the answer analytically?

Completing the Square Again

If I expand the bracket in the formula for the area, $A = w(100 - 2w)$, I get:

$$A = 100w - 2w^2$$

This is a quadratic, so I should be able to complete the square with it, but I want to rearrange it a smidge first:

$$A = -2w^2 + 100w$$
$$A = -2(w^2 - 50w)$$

This has made the stuff in the brackets look more like the examples in the previous chapter:

$$w^2 - 50w = (w - 25)^2 + c$$

I've written '+ c' here because I know there is some number I need to work out to make the equation true.

$$(w - 25)^2 = (w - 25)(w - 25) = w^2 - 50w + 625$$

Squaring the bracket gives me an extra 625 on the right-hand side of the equation that I don't want, but if I make $c = -625$ this compensates:

$$w^2 - 50w = (w - 25)^2 - 625$$

If I put this back into our area equation (using some square brackets to keep it clearer) and rearrange a bit:

$$A = -2[w^2 - 50w]$$
$$A = -2[(w - 25)^2 - 625]$$
$$A = -2(w - 25)^2 + 1250$$

Now the magic can happen. When you put a quadratic in completed square form, it tells you about its maximum or minimum point. The area formula is made of a negative part, $-2(w - 25)^2$, and a positive part, 1250. Therefore, to maximize the area I need to minimize the negative part as there is no way to increase the positive part. As the bracket is squared and all square numbers are positive, the smallest I can make the bracket, when squared, is zero, which happens when $w = 25$. This shows me that the most the area can be is 1250 yd^2 when $w = 25$ yd. So the knight's son can get the most land with a rectangle 25 yards wide and 50 yards long.

This process is tricky, but hopefully you can see the logic in the final conclusion. Completing the square only works for quadratic equations – what if the formula I want to optimize uses powers higher than two?

Calculus

An area of mathematics called *calculus* enables us to look at how values change. The process of *differentiation* allows us to work out the rate of change of a value and can show us the maximum and minimum points without needing to draw the graph.

Key to differentiation is the concept of gradient and what the gradient is at maximum and minimum points. Imagine you are on a walking holiday in the Mathematical Highlands, where all the hills are lovely smooth curves. Much of your time will be spent walking uphill – up a positive gradient. Also, you'll spend time walking downhill where the gradient is negative in relation to your direction of travel. At the mountaintops and the valley floor – i.e. the maximum and minimum altitudes in your journey – you will be on level ground where it changes from rising to falling or vice versa. Differentiation allows us to find these points of zero gradient.

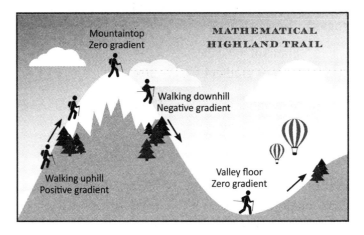

Exactly why and how differentiation works is beyond the scope of this book – grab your nearest A Level textbook for more detail.

So, today you have planned a 10 km hike through some cubic mountains. Cubic equations are ones with an x^3 term in them, and our particular mountains are described by the formula below, where x is your distance horizontally from the starting point and y is the distance vertically:

$$y = 2x^3 - 15x^2 + 24x + 10$$

We're going to use differentiation to work out where the summit of the mountain is and also the valley floor.

Differentiation, in its simplest form, changes an equation into the equation of the gradient. To do this, I modify the x terms using the following rule:

$$x^n \text{ becomes } nx^{n-1}$$

So our x^3 becomes $3x^2$ using this template and x^2 becomes 2x. Remember, x is x^1 and so becomes $1x^0$ when we differentiate. The 10, which has no x involved, becomes zero. If I do this to the whole equation (remembering that $x^0 = 1$, see page 61):

$$\text{gradient} = 2 \times 3x^2 - 15 \times 2x^1 + 24 \times x^0 + 0$$
$$\text{gradient} = 6x^2 - 30x + 24$$

Now, remember that the peak and the valley floor will be where the gradient is zero:

$$6x^2 - 30x + 24 = 0$$

This is one of our old friends, the quadratics, which we have multiple ways to solve (see page 99). Whichever way you do it, you should find that the equation is true when x = 1 and 4. How can we tell which is the peak and which is the valley floor? If I substitute these values into the original equation, I can compare the results to see which is highest:

$$y = \quad 2x^3 \quad -15x^2 \quad +24x \quad +10$$

if x = 1, $\quad y = \quad 2 \times 1^3 \quad -15 \times 1^2 \quad +24 \times 1 \quad +10$

$\quad\quad\quad\quad y = \quad\quad 2 \quad\quad -15 \quad\quad +24 \quad\quad +10$

$\quad\quad\quad\quad y = \quad\quad 21$

if x = 4, $\quad y = \quad 2 \times 4^3 \quad -15 \times 4^2 \quad +24 \times 4 \quad +10$

$\quad\quad\quad\quad y = \quad\quad 128 \quad\quad -240 \quad\quad +96 \quad\quad +10$

$\quad\quad\quad\quad y = \quad\quad -6$

So we can see that the first zero-gradient place is the peak and the second is the valley floor.

Newton versus Leibniz[†]

Calculus has many uses other than planning walking holidays, and since its invention in the 1600s it has become one of the most crucial areas of mathematics, allowing many problems to be solved. Back in the day, though, two giants of the mathematical world had an argument over who came up with it first that created a long-lasting split between British and European mathematics.

In the British corner – a gentleman who needs no introduction – Sir Isaac Newton. Watcher of falling apples, alchemist, Cambridge professor, Member of Parliament, Master of the Royal Mint and President of the Royal Society.

In the German corner – Gottfried Leibniz. Child prodigy, philosopher, diplomat, inventor of mechanical calculators, proponent of binary and eternal optimist.

Both of these gentlemen independently invented calculus at around the same time. Leibniz published his ideas first, but Newton claimed that Leibniz nicked the idea from some papers he waved around at the Royal Society, where Leibniz was a member. The conflict, being as it was between two high-profile academics, consisted mostly of strongly worded letters and the distribution of lampooning pamphlets.

Leibniz died before the controversy was concluded. Today, it seems clear that both gentlemen had the mathematical prowess to invent calculus and they both would have read various other mathematicians' work as inspiration. Suffice to say, Leibniz's notation is the one that has survived until today.

[†] Fig Newtons and Choco Leibniz are two of my favourite biscuits. Choco Leibniz *are* named after the mathematician and are made in Hanover where he lived and worked. Fig Newtons, however, are named after a town in the USA.

Chapter 16

ALGORITHMS

Algorithms are inextricably linked with computer science, but their origins are much older than computers. An algorithm is a set of instructions. You take an input – usually a number or numbers that you choose – and follow a set of procedures to receive an output. Although algorithms are not necessarily part of algebra, I've included them here as most of them use the idea of variables as their inputs.

Russian Peasants

This example is sometimes called the Russian peasant algorithm, although there is evidence of its use way before there was a Russia to have peasants. If I wanted to multiply 35 by 47, the algorithm says to take the smaller number and keep halving it, ignoring any remainders:

35

17

8

4

2

1

Next step is to make another column using the larger number, doubling each time:

35	47
17	94
8	188
4	376
2	752
1	1504

Then cross out all the rows with an even number in the left-hand column:

35	47
17	94
~~8~~	~~188~~
~~4~~	~~376~~
~~2~~	~~752~~
1	1504

Add up whatever remains in the right-hand column to get the answer:

$$47 + 94 + 1504 = 1645$$
$$\text{Therefore, } 35 \times 47 = 1645.$$

Ta-da! This algorithm is a way of multiplying two numbers together despite only being able to double, halve and add numbers – ideal, perhaps, for a Russian peasant who doesn't know their times tables. By breaking a complex task into a simple set of procedures, algorithms are also ideal for programming into computers. In fact, most computer programmers spend their time teaching computers algorithms to achieve certain outcomes, whether that is multiplying numbers, ordering a list or filtering your selfies to make you look younger.

Computer Programs

The first real computer program was an algorithm to compute a series of numbers called the Bernouilli numbers (after Jacob Bernouilli, see page 74), which are difficult to calculate by hand. What is amazing about it is that the program was created long before the computer to use it was built. Ada Lovelace (1815–52) was the extraordinary Englishwoman who came up with it while working with her compatriot, the engineer Charles Babbage (1791–1871) on the design of his never-to-be-made steam-powered computer, the Analytical Engine. She also saw beyond the Engine as a mere calculating machine and realized that anything that could be encoded numerically could be acted on by it, including music, pictures and letters, presaging exactly what we enjoy with our computers today.

As electronic computers were invented and made, they could be used to work on problems that would take a vast amount of time for humans to solve. A good example of this is Alan Turing's *bombe* – a computer that the team at Bletchley Park used during the Second World War to decrypt Nazi messages enciphered on the Enigma machine.

The concepts involved are relatively simple. The Enigma was like a supercharged code wheel – you typed a letter from your message on your keyboard, and the machine would pass your letter through three code wheels (called *rotors*) and light up the enciphered version of your letter. The tricky bit was that the rotors could rotate after every letter, effectively meaning that every letter of your message would be encrypted using a different cipher. Unless you knew the initial starting position of each rotor, there was little hope of reading the message.

The bombe exploited the only flaw in this system, which was that a letter would never be encrypted back into itself. The bombe would work out the starting positions of the rotors by trying each combination to see whether it resulted in a letter being encrypted as itself, which would eliminate that particular setting. A straightforward algorithm but, owing to the number of possible starting positions, use of different rotors and some other tricks on the Enigma machine, there were 159 quintillion settings. A perfect job for a computer that can work flawlessly, many times faster than a person.

Today, algorithms run behind the scenes in many areas we take for granted. When you get directions from your location to where you want to go, the computer uses an algorithm that works out the shortest (in either distance or time) route to take you there, doing many calculations to optimize the result for you. One such shortest-path algorithm is called Dijkstra's algorithm, after the Dutch computer scientist Edsger Dijkstra (1930–2002). Any time you send information over the internet, an algorithm is used to keep it safe (see page 47). When you get a spreadsheet to alphabetize a list or sort some numbers, a sorting algorithm is used. Whenever you talk to someone on the telephone or over the internet, a Fourier transform algorithm digitizes the sounds and images, named after the Frenchman Joseph Fourier (1768–1830), who did a lot of work on the mathematics of vibration when he wasn't discovering the greenhouse effect. Link analysis is a technique that uses various algorithms to establish links between data. This influences many areas, from search engines choosing the most appropriate results for you, to social media apps deciding which ads to show you, to marketing and medical research and police forces collating and cross-referencing evidence.

The work of computer scientists and mathematicians is not done, however – there are still a few problems that do not have an efficient algorithm to solve them.

Bin Packing and Travelling Salespeople

Getting stuff from one place to another efficiently and cheaply is pretty much the cornerstone of modern retail, especially as we increasingly buy things online and have them delivered to our doors. So it is quite annoying for retailers that there are no quick algorithms that find the best way to pack a delivery truck, nor to give the driver the shortest route.

The bin packing problem, as it is unglamorously known, concerns trying to find the best way to fit a number of packages of known volume into a given space – e.g. parcels into the back of a delivery lorry. The problem here is that there is no simple way to decide which is the best approach to take – largest first? Smallest first? Or some other criterion? It is possible to analyse every possible combination and choose the best, but often this calculation would take far too long.

Assuming you manage to pack your van in some sort of sensible fashion, how do you determine the best route to take to deliver all the parcels?

This problem was originally called the travelling salesman problem when it was first considered by the Irish mathematician William Rowan Hamilton (1805–65) and the British clergyman Thomas Kirkman (1806–95). They considered a salesman who needed to travel around certain towns and get back to his factory afterwards by the shortest route without revisiting any town. Hamilton even produced a board game that illustrated the problem, called the Icosian Game, in 1857, which you can still get as an app today.

It was quickly realized that existing algorithms would not necessarily produce the correct route. For instance, the nearest neighbour algorithm, as the name suggests, moves you to the closest unvisited town, but this can leave you with a very long way to go later on.

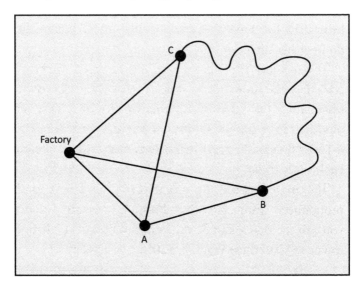

In the example above, the nearest neighbour algorithm would take you from the factory to A, then B, leaving you no choice but to go to C via the long and winding road. Factory – C – A – B – factory would be much shorter.

So the only way to work out the shortest route is to list all of them. In our example, with three towns to visit, this is not too bad:

> factory – A – B – C – factory
> factory – A – C – B – factory
> factory – B – A – C – factory
> factory – B – C – A – factory
> factory – C – A – B – factory
> factory – C – B – A – factory

If you think about it, some of the routes are reverse copies of each other, such as the first and the last in the list, so we don't count these as being separate routes. This brings the list down to three possible routes that I could do in either direction.

To calculate the number of routes more directly, I consider that from the factory I have three choices of town, then two unvisited towns from there, then one unvisited town and finally back to the factory. This is $3 \times 2 \times 1 = 6$ choices of route, then halved due to the forwards–backwards thing.

The mathematical shorthand for $3 \times 2 \times 1$ is $3!$, pronounced 'three factorial'. From the example above, you can see that I had $3! \div 2$ routes to check. If I had 20 towns to visit, there would be $20! \div 2$ to check:

$$\frac{20!}{2} = \frac{20 \times 19 \times 18 \ldots \times 3 \times 2 \times 1}{2} \approx 1{,}216{,}451{,}000{,}000{,}000{,}000$$

This figure is in the same ballpark as the number of settings on the Enigma machine. As you can see, factorial values increase very rapidly. Mathematicians call this sort of problem NP, which stands for *nondeterministic polynomial time*, jargon meaning that as the number of things you are considering increases, the number of combinations increases as a power of that, drastically increasing the number of combinations that need to be considered and thus the time needed to work out the best answer.

There are a number of algorithms that give something close to the best result in a reasonable amount of time. These *heuristic* algorithms are similar to what we do when we look at a map and plan a route ourselves – we don't know that we have found the perfect route, but it works and didn't take too long to figure out.

If you were to go away and come up with an efficient algorithm – one that finds the best solution without looking at every possible combination – I can think of quite a few companies that would hand over a lot of money for the right to use it.

Chapter 17

FORMULAE

There are formulae everywhere in this book. They are such a fundamental part of using mathematics that it would have been hard to avoid them. In this section, I want to take a closer look at some more and think about how they work.

Rearranging Formulae

The formula for converting a temperature in Celsius (shown by C) to Fahrenheit (shown by F) is as follows:

$$F = \frac{9C}{5} + 32$$

At the moment, F is the *subject* of the formula as it is the output you get when you input a value for C. What if we want to convert temperatures the other way, from Fahrenheit to Celsius? This is particularly useful if you have any old cookery books.

Changing the subject of a formula is very similar to solving an equation. I go through the same process, but rather than getting a value at the end as I did with the equations I solved, I get a rearranged formula:

$$F = \frac{9C}{5} + 32$$

$$(-32)$$

$$F - 32 = \frac{9C}{5}$$

$$(\times 5)$$

$$5(F - 32) = 9C$$

$$(\div 9)$$

$$\frac{5(F - 32)}{9} = C$$

Our new formula is the *inverse* of the first one because they do the opposite of each other. So, for instance, if I say C = 25° Celsius:

$$F = \frac{9 \times 25}{5} + 32 = 77$$

So I know that 25°C is 77°F. If I then put this through my new formula:

$$C = \frac{5(77 - 32)}{9} = 25$$

77°F is 25°C – back where we started.

Better Late than Never

When the American mathematician George Dantzig (1914–2005) was at university he was late for one of his lectures. He saw two problems written on the blackboard and assumed they were the assignment set by the lecturer. He duly handed in the work and was very surprised when he was told by the lecturer that they weren't an assignment but, in fact, two unsolved problems in mathematics. He was then told that these solutions would suffice for his PhD thesis.

Sequences

One of the things that mathematicians try to do is to go from the specific to the general. In the example above, 25°C being equivalent to 77°F is a specific example, whereas the formulae allow me to work out any general value I want.

Mathematicians are often interested in number patterns. Here's a really simple one:

$$2, 4, 6, 8, 10, 12, 14, 16, 18, 20 \ldots$$

These are the even numbers, or the multiples of two if you prefer. You could keep on going with the sequence without any trouble as we recognize that the way to get the next number is to add on two. Each number in the sequence is called a *term*. The first term in the sequence is two, which a mathematician would write as $t_1 = 2$. I've chosen t for term, but I could have chosen any other symbol. The second term is 4, so $t_2 = 4$. From here, $t_3 = 6$ and $t_4 = 8$.

Now to go from the specific to the general. I said earlier that to find the next term in the sequence, I need to add on two. If I am currently on the nth term of the sequence, t_n, I can find the next term with the formula:

$$\text{next term} = \text{current term} + 2$$
$$t_{n+1} = t_n + 2$$

This is called the *inductive* definition of the sequence. The sequence it gives, however, depends on what you set as the first term. If $t_1 = 2$ I get the even numbers as we've seen, but if I say $t_1 = 1$, I get 1, 3, 5, 7, etc. – i.e. the odd numbers, instead.

A very famous sequence is the Fibonacci numbers:

$$1, 1, 2, 3, 5, 8, 13, 21, 34 \ldots$$

These are named after Fibonacci (see page 23), and the rule is that to get the next term you must add the previous two terms. Written inductively, this is:

$$t_{n+1} = t_{n-1} + t_n \text{ with } t_1 = 1 \text{ and } t_2 = 1$$

These numbers appear in nature a lot, especially in plants. The numbers of shoots on stems and roots often form Fibonacci numbers, as well as the spiral patterns in the seed heads of flowers and on pineapples and pinecones.

While these inductive formulae are useful for describing the sequence in general terms, we can't use them to calculate later terms without working out all the ones in between. What if I want to know the 1000th term, though? For this I need an nth term formula.

Looking back at the even numbers, we can see a pattern between the term number and the term itself. $t_1 = 2$, $t_2 = 4$, $t_3 = 6$, $t_4 = 8$: the term is always double the term number:

$$t_n = 2n$$

This formula allows me to calculate the 1000th term directly: $t_{1000} = 2 \times 1000 = 2000$. Can we work out an nth term formula for the Fibonacci numbers?

Well, there is one, but it's slightly more complicated. Our even number sequence is relatively easy to work out because the sequence increases by the same amount every term. The Fibonacci numbers increase by different amounts each time.

Back to φ

We saw on page 90 that φ = 1.61803398 ... What does this have to do with the Fibonacci numbers? Well, if you divide consecutive terms of the Fibonacci sequence, we see that:

$$1 \div 1 = 1$$
$$2 \div 1 = 2$$
$$3 \div 2 = 1.5$$
$$5 \div 3 = 1.66666 \ldots$$
$$8 \div 5 = 1.6$$
$$13 \div 8 = 1.625$$
$$21 \div 13 = 1.61538 \ldots$$
$$34 \div 21 = 1.61905 \ldots$$
$$55 \div 34 = 1.61764 \ldots$$

The values obtained are hovering around φ. If I skip ahead a bit in the Fibonacci numbers to the nineteenth and twentieth values:

$$6765 \div 4181 = 1.61803396 \ldots$$

This is the same as φ to seven decimal places, so it seems clear that there is a relationship between the Fibonacci numbers and the golden ratio.

To explain the value of φ, we can look at that beautiful rectangle again – if I assume that the long side of the rectangle is of length φ and the shorter side is of length one, then I can mark it out as follows:

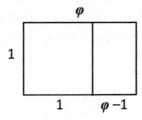

One of the rules of the golden ratio is that the smaller rectangle must have the same proportions. The big rectangle has sides of length φ and 1. The smaller rectangle has sides of length 1 and $\varphi -1$. So if I divide the length of each rectangle by its width, I should get equivalent fractions:

$$\frac{\varphi}{1} = \frac{1}{\varphi -1}$$

Anything divided by one is itself, so:

$$\varphi = \frac{1}{\varphi -1}$$

This is an equation that I can solve. I need to get all the φs together, so I'll multiply both sides by the denominator of the fraction:

$$\varphi (\varphi -1) = 1$$

Multiply out the brackets:

$$\varphi^2 - \varphi = 1$$

Ooh, a quadratic equation (see page 99). I need to make it equal to zero to solve, so I'll subtract one from both sides:

$$\varphi^2 - \varphi - 1 = 0$$

This quadratic won't factorise nicely, so I'm left with completing the square or using the quadratic formula. As we're in the Formulae section, I'll choose the latter:

$$\varphi = \frac{-(-1) \pm \sqrt{(-1)^2 - 4 \times 1 \times -1}}{2 \times 1}$$

$$\varphi = \frac{1 \pm \sqrt{5}}{2}$$

This gives me φ = 1.6180339887 ... and –0.6180339887 ... I reject the negative result as I can't have a rectangle with a negative length.

So, back to our formula for the nth Fibonacci number. Abraham de Moivre (1667–1754) was a French Protestant who moved to London to escape persecution. While in London, he met Isaac Newton and the two became friends – Abraham even became Isaac's go-to guy if he got stuck on some maths! Anyway, de Moivre first published this formula for the Fibonacci numbers:

$$F_n = \frac{\varphi^n - (1 - \varphi)^n}{\sqrt{5}}$$

What is amazing about this is that the formula uses two irrational numbers (φ and $\sqrt{5}$) to produce an integer Fibonacci number.

De Moivre's Prophecy

De Moivre lived to the ripe old age of eighty-seven, a very good innings in those days. However, legend has it that he began to notice that he was sleeping for longer each night and predicted that he would die when he required more sleep than there were hours in the day. He calculated the date when this would occur and died on this exact day.

The Three-Body Problem

Understanding formulae is crucial if you want to be good at astronomy, which many mathematicians have studied over the ages. Although the problem had been considered before, it was Isaac Newton who first formally stated the three-body problem in his magnum opus *Principia Mathematica*. The problem considers three objects or *bodies*, their motion in space and how their gravity affects each other. We saw the formula underpinning all this in our discussion of powers (see page 66), but here it is again:

$$F = \frac{Gm_1m_2}{r^2}$$

Newton's law of universal gravitation, as it is known, says that the force of gravity between any two objects is equal to the product of their masses ($m_1 \times m_2$) divided by the square of the distance between them (r^2), multiplied by the *gravitational constant* (G). It turns out that it is fairly easy to work out the motion of two bodies, like the earth going around the sun. If you add a third body, say the moon orbiting the earth, the maths becomes much, much trickier. As you keep on adding bodies to the system, the governing equations become so complex that it can be impossible to solve them analytically. This means that you can't solve the equations using algebra, but you may be able to solve them numerically by using techniques to find numbers that fit the equations. This is usually very time-consuming and these days is something we'd get computers to do for us.

The upshot is that the orbits of things in our solar system, be they planets, moons, asteroids, satellites or whatever, wobble a bit due to the gravity forces from the other bodies in the solar system changing as things move closer or further away.

Astronomers have become good at looking at the wobbles of the planets and using them to deduce the presence of other bodies in the solar system. In 1846 the Frenchman Urbain Le Verrier (1811–77) of the Paris Observatory, after months of calculation, deduced the existence of Neptune from the difference in Uranus' actual orbit from the one predicted by the current model of the solar system. His calculations were then used to observe the planet. This mathematical discovery of Neptune is considered one of the greatest scientific feats of the nineteenth century.

Interestingly, Le Verrier went on to look at Mercury's wobbles, which predicted the presence of another planet, which was dubbed *Vulcan*. Nobody could find it, though, and it wasn't until 1916 that Albert Einstein proposed that the wobble was due to the effects of *general relativity* from the proximity of the sun.

Anyway, astronomers continued to observe Neptune, ran the numbers and discovered that it too was being affected by another body that had not been discovered yet. It remained elusive until the American Clyde Tombaugh (1906–97) discovered Pluto in 1930 by comparing telescope pictures of the night sky taken two weeks apart. It was tiny, and its size couldn't account for the perturbation of the orbits of Neptune and Uranus, which was eventually discovered to be due to overestimating the mass of Neptune.

Since the discovery of Pluto, it has been pushed down to seventeenth place in the solar system according to mass – even our own moon is heavier than it. In 2006 the International Astronomical Union provided a formal definition of a planet which Pluto did not quite fulfil. Hence, it has been downgraded to the status of a dwarf planet along with several other contenders such as Eris and Sedna.

This didn't stop some of Clyde Tombaugh's ashes being carried aboard the New Horizons spacecraft which reached Pluto in 2015.

Sedna has a very strange orbit, and mathematical analysis of this and several other distant dwarf planets has led some scientists to believe in the existence of Planet Nine, four times larger than the earth, with an orbit at right angles to the rest of the solar system. The numbers work – all we have to do now is find it!

Noether and Mirzakhani

Mathematics is a field historically dominated by men. I have no doubt that this is due to inequality in education and opportunity rather than some neurological difference between the sexes. Fortunately, this is slowly changing, but it does mean that any women who have made a name for themselves in mathematics must have been truly exceptional.

Emmy Noether (1882–1935) was a German who is considered by many (including Einstein) to be one of the best mathematicians, male or female, ever. Despite her ability and credentials, she was not paid for teaching at universities, and had to lecture under male colleagues' names as an assistant. While teaching, she developed what is now called Noether's theorem, which has far-reaching implications for physics.

Maryam Mirzakhani (1977–2017) was an Iranian mathematician, notable for being the first female recipient of the Fields Medal, the highest award in mathematics (as there is no Nobel Prize for mathematics). After undergraduate work in Iran, she moved to the USA, where her work in geometry earned her high renown. She died from cancer at the age of forty and is one of very few women to have their picture published in the Iranian press with her hair uncovered.

4

GEOMETRY

Chapter 18

AREA AND PERIMETER

Geometry – ancient Greek for 'earth measurement' – has been one of the most useful areas of mathematics for a very long time. We often associate ancient cultures with the monuments they left behind, and geometry was what they used to design them.

Here's a quick refresher of some basic geometry.

Angle Facts

We usually name lines on a diagram after the points at the ends of the line. For instance, if two lines, AB and CD cross at point E, I could represent this as below:

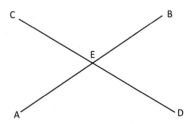

This creates four angles at E. To describe an angle, you

use the two lines that make it. So to describe the angle marked below, I see that it is the angle made when I go from A to E and then to C. Hence it is called angle AEC, or ∠AEC. ∠CEA is the same thing.

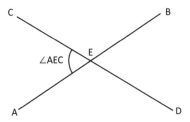

Owing to the base-60 number system of the ancient Mesopotamians, we have 360 degrees around a point. A straight line effectively splits a point or vertex (such as E) in two, giving 180° on each side. Across a vertex, we see angles that are equal: ∠AEC and ∠BED. We call pairs of angles like this vertically opposite, with 'vertically' being the adverb from 'vertex', rather than from 'vertical'.

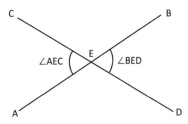

We don't know the value of the angles in my diagram, but I can say that ∠AEC and ∠BED are both *acute* – less than 90° – while ∠CEB and ∠AED are between 90° and 180°, making them *obtuse*. Angles larger than 180° are called *reflex* angles.

Parallel Line Facts

Parallel lines are always the same distance apart and never meet. If we draw a line across the parallel ones (shown with chevrons), we can see some more relationships:

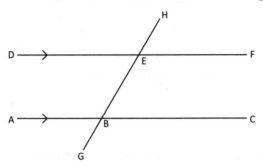

The line GH crosses the parallel lines DF and AC at the same angles, producing two identical intersections at B and E. Angles that occupy the same position in each intersection are equal and are known as *corresponding* angles, such as ∠HEF and ∠HBC:

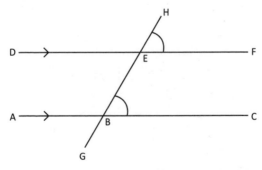

We can see that ∠HEF and ∠BED are vertically opposite and so must be equal. This, in turn, means that ∠BED and ∠HBC must be equal and are said to be *alternate*.

It's All Greek to Me

What started as a practical, engineering-based field of study soon became a field of theoretical study in its own right, especially once the philosophically minded Greeks got their hands on it. Thales (*c.* 624–526 BCE), Pythagoras (*c.* 570–495 BCE) and Eudoxus (*c.* 390–337 BCE) all made massive contributions, but it was Euclid (fl. *c.* 300 BCE), a Greek who lived in occupied Egypt, who has been named the Father of Geometry.

Why him and not the others? The answer is that he wrote a bestseller.

His *Elements* is considered to be the greatest textbook ever written and until the twentieth century it was standard reading for anyone who wanted to be considered an intellectual. Today, much of its content is dispersed through the mathematics you learn at school. Mathematicians hail it as one of the first examples of rigorous, thorough proofs that set the benchmark for everything since.

Euclid started with things that were obviously true and did not need proving, which he called *axioms*. For instance, that you can join any two points with a straight line, or that you can extend any straight line if you want to. He then used these axioms to demonstrate geometrical and number theorems, proving them as he went along.

As an example, Euclid showed that the angles in a triangle must have a sum of 180° using only the ideas of corresponding and alternate angles. Imagine a triangle, ABC, drawn between parallel lines:

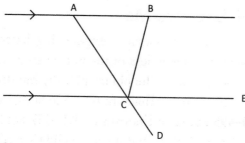

Euclid said:

∠ABC and ∠BCE are alternate and therefore equal.

∠BAC and ∠ECD are corresponding and therefore equal.

Therefore, ∠ABC + ∠BAC = ∠BCE + ∠ECD:

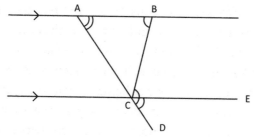

If we add the third angle of the triangle, ∠ACB, to both sides of this equation, we get:

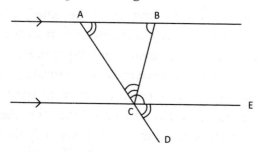

∠ABC + ∠BAC + ∠ACB = ∠BCE + ∠ECD + ∠ACB

angles in the triangle = angles on a straight line

Therefore, the angles in a triangle add up to 180°.

Perimeter

Perimeter can refer to two things – the distance around the outside of a two-dimensional shape, or the outside of the shape itself. For instance, I could say that the perimeter of some land was 210 metres long, or I could say that the perimeter is rectangular in shape. Ever since the first hunter-gatherers settled down into farming, people have been interested in shapes. We have evidence from several ancient cultures of maths problems relating to the size and shape of fields, quite often so the farmers could be taxed.

The perimeter of any straight-sided shape or *polygon* is straightforward to measure. It becomes less straightforward when you introduce curves or *arcs*. It is possible to get something flexible like rope or string to lie on a curve and you can then straighten it to measure the length of the arc. However, mathematicians like formulae and so the quest for a formula for the perimeter of a circle occupied ancient mathematicians for some time.

Circles

To work out the perime-ter of a circle, called the *circumference*, ancient mathematicians noticed a relationship between the side of a square and the circumference of the circle with a radius equal to the side of the square:

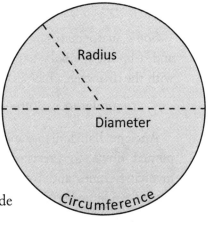

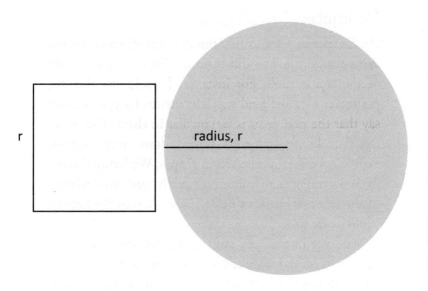

The perimeter of the circle is six-and-a-bit times as long as the side of the square:

circumference = six-and-a-bit × radius

Some mathematicians have proposed giving the six-and-a-bit the symbol τ, tau, the Greek letter t. If you work with the diameter, which is twice the radius, you get:

circumference = three-and-a-bit × diameter

Ancient Babylonians and Egyptians had the value pinned down to precision sufficient to build things involving circles and arcs, but it was the Greek maths wizard Archimedes (*c.* 287 to *c.* 212 BCE) who first came up with a systematic way of working out values of this number with increasing precision.

His method was to approximate a circle using two hexagons as below:

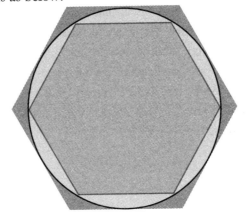

It is easy to measure the perimeter of the hexagons as their sides are straight. Archimedes reasoned that the circumference must be somewhere between the perimeter of the inner, or *inscribed*, hexagon and the perimeter of the outer, or *circumscribed*, hexagon. Archimedes then doubled the number of sides of the polygons to get increasingly accurate approximations of the circumference. When he reached 96 sides, his results told him that the circumference was approximately $\frac{22}{7}$ times the diameter.

This value was known as Archimedes' constant for a long time, but the symbol π came into use later on, being the first letter of the Greek word for periphery. Archimedes' value is very close to the actual value and is still used as an approximation in non-calculator maths exams to this day.

So for a circle of radius r or diameter d we get:

$$\text{circumference} = \pi d = 2\pi r$$

Subsequent work has shown that π is irrational (see page 14), so the best we can do is approximate it. π has also been shown to be *transcendental*. This has nothing to do with spirituality – it means a number that is not the solution of an equation where the coefficients and powers of the terms are integers. It can be pretty hard to show that a number is transcendental, and was first shown for π in 1882 by the German mathematician Ferdinand von Lindemann (1852–1939).

Area

Area is the amount of flat space a shape occupies. Von Lindemann's work showed that it was not possible to *square the circle*, which is to draw a square with the same area as a circle using only a pair of compasses and a straight edge, something that geometers had been trying to do since antiquity.

We spend money in proportion to area quite often when we buy paint for walls or office space. Key to working out the area of polygons is being able to work out the area of a rectangle (length × width) and a triangle ($\frac{1}{2}$× base × height). From here, you can split any polygon into rectangles and triangles to work out its area. For example:

Here, the *quadrilateral* (four-sided polygon) can be

split into three triangles and a rectangle. However, if we wish to know a curved area, this becomes trickier. We saw the formula for area of a circle on pages 76–77, but how could I work it out from scratch? What if I tried splitting my triangle into bits and rearranging it?

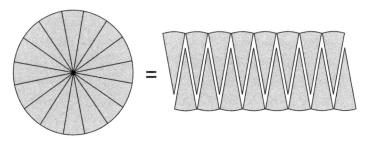

Here I've cut the circle into equal slices and rearranged them into something that approximates a rectangle. The more slices, the more rectangular my rearrangement would become until, eventually, my slices would effectively be very thin rectangles themselves. We know that the area of a rectangle is its width × length. The width of my rectangle is the distance from the centre of the circle to the edge – the radius. The length of the rectangle must be half the circumference, as I have half the 'crusts' on each side of the rectangle. So the area of the rectangle is radius × (circumference ÷ 2):

$$\text{area of circle} = \text{radius} \times (\text{circumference} \div 2)$$
$$= r \times (2\pi r \div 2)$$
$$= r \times \pi r$$
$$= \pi r^2$$

My local pizza restaurant produces several different sizes of pizza, which it sells according to diameter. A small pizza has a diameter of 9.5 inches and costs £13.99. A personal pizza has a diameter of 7 inches and costs £6.99 – which offers the most pizza for my pennies?

Well, if I compare their areas to one decimal place:

$$Area_{small} = \pi \times \left(\frac{9.5}{2}\right)^2 = 70.9 \text{ square inches}$$

$$Area_{personal} = \pi \times \left(\frac{7}{2}\right)^2 = 38.5 \text{ square inches}$$

I can now see that two personal pizzas will have an area of 38.5 × 2 = 77 square inches for £13.98, so are better value for money than one small. Check your own pizzeria before you buy!

Calculus Again

Being able to work out the area of circles helps us to find the area of shapes made with bits of circles. Shapes made with other curves can be found using *integration*, the opposite of differentiation (see page 111). If we can describe the curve using an equation, we can integrate it to get another equation which tells us the area under the curve.

Calculus can also be used in three dimensions to work out the volume of a curved shape like a sphere. First, however, no geometry section would be complete without a look at Pythagoras.

Fractals – Finite Area, Infinite Perimeter

Here is an algorithm for creating a shape. First, start with a square:

Then add another square to the middle of each edge:

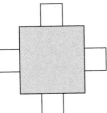

Repeat ad nauseam:

These self-repeating patterns are known as *fractals* and they exhibit some very interesting properties.

If I double the lengths of the edges in a shape, we call this an *enlargement with scale factor 2*. Ordinarily, the area will increase by the square of the scale factor, so in this case by $2^2 = 4$ times (see the tiles on page 86 to visualize this). Does this hold for the fractal above?

The first time I add more squares (called an *iteration*), the length of each side increases by a scale factor of $\frac{5}{3}$. You can see this if you imagine that the original side of the square was three units long and has been replaced by a shape that is five units long:

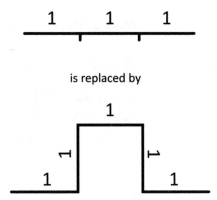

is replaced by

The area changes from 9 (3 × 3) to 13. Now, if the usual rules of enlargement hold, we should find that:

$$9 \times \left(\frac{5}{3}\right)^2 = 13$$

i.e. the area is increased by the scale factor squared. The left-hand side of this calculation actually gives an answer of 25. Each iteration we perform increases the length of the edges of the shape by more than we'd expect from the increase in area.

The upshot of this is that if I continue the iterations indefinitely, the perimeter of the shape produced tends to infinity while the area remains finite. Something similar happens if we work in three dimensions – we get solids that have a finite volume but an infinite surface area. Many plants and animals exploit this wherever surface area is a key requirement, such as in lungs and leaves.

Chapter 19

PYTHAGORAS' THEOREM

The theorem named after Pythagoras (*c.* 570 to *c.* 495 BCE) was not invented by the man himself. Records from various cultures around the world, predating Pythagoras, show it being used. Pythagoras, however, was among the first people to prove the theorem.

Pythagoras' theorem concerns itself with right-angled triangles:

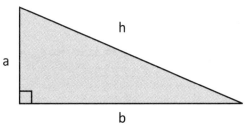

I have labelled the two sides at right angles to each other a and b. The third side I have labelled h for *hypotenuse*, the name given to the longest side of a right-angled triangle, which is always opposite the right angle itself. Pythagoras' proof is particularly elegant because it is done purely by rearranging triangles.

I can take four of these triangles and arrange them to form two squares of equal area:

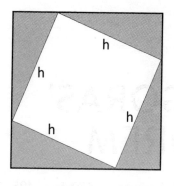
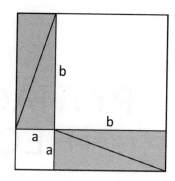

Pythagoras said that the internal square on the left must have an area of h × h = h². The two squares on the right must have areas of a² and b². As each large square has the same area and the triangles have not changed in size, it must be true that:

$$h^2 = a^2 + b^2$$

Pythagoras' theorem allows us to calculate an unknown side on a right-angled triangle, as in the examples below:

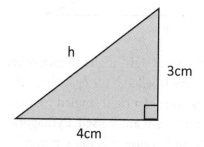

$h^2 = a^2 + b^2$

$h = \sqrt{a^2 + b^2}$

$h = \sqrt{3^2 + 4^2}$

$h = \sqrt{9 + 16}$

$h = \sqrt{25}$

$h = 5$ cm

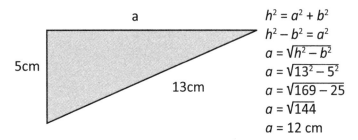

$$h^2 = a^2 + b^2$$
$$h^2 - b^2 = a^2$$
$$a = \sqrt{h^2 - b^2}$$
$$a = \sqrt{13^2 - 5^2}$$
$$a = \sqrt{169 - 25}$$
$$a = \sqrt{144}$$
$$a = 12 \text{ cm}$$

In both these examples the three sides of the triangles are integers. Integers that work in Pythagoras' theorem are called *Pythagorean triples*. 7, 24 and 25 and 8, 15 and 17 are other examples.

Fermat's Last Theorem

Diophantus (c. 210 to c. 290) was a Greek mathematician who lived in Egypt. In his series of books called *Arithmetica*, he considers equations that look similar to Pythagoras' theorem. Pierre de Fermat (1607–65), a French lawyer who killed time doing mathematics, wrote a note in his copy of the *Arithmetica*, saying that he had a 'truly marvellous proof' that equations of the form $x^n + y^n = z^n$ only work with integers when n is two, i.e. the Pythagorean triples. The proof was never discovered and it took the best part of 400 years before Andrew Wiles (b. 1953), a British mathematician, proved it conclusively. In doing so, he also paved the way for the proof of the *modularity theorem*, which was in the *Guinness Book of Records* as having the most failed attempted proofs.

What Fermat's own proof might have been, no one knows.

Why the big deal about Pythagoras' theorem? Well, it turns out that working out lengths of hypotenuses is very helpful in working out distances between two points in *coordinate geometry*.

From A to B

Descartes' (see page 107) x-y coordinate system helps us describe positions, lines and shapes. Imagine if I have two points: A at position (1,2) and B at position (5,5), as shown:

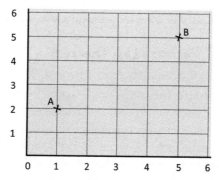

How can I find the exact distance between these points? By forming a right-angled triangle:

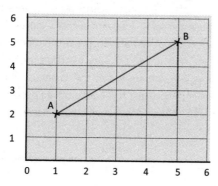

I can see that the base of the triangle is 4 units and the height is 3 units. This is the same as the triangle on page 149, so we know the distance AB must be 5 units.

Pythagoras' theorem also works in three dimensions. Imagine I have a box with width x, length y and height z and I want to work out the length of its diagonal:

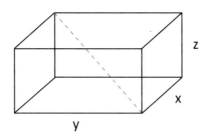

I can use Pythagoras' theorem twice to find it. First I make a right-angled triangle:

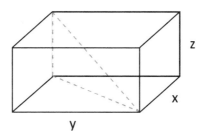

The diagonal I want is the hypotenuse of this triangle. I know that the height of the triangle is z, but I don't know the base yet. If I look at the rectangle that makes the bottom of the box, however, I can see a way to find that distance using Pythagoras' theorem:

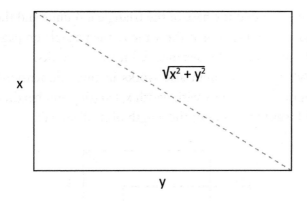

Now that I know the base and height of the triangle, I can work out the length of the diagonal, which I'll call d:

$$d^2 = \left(\sqrt{x^2 + y^2}\right)^2 + z^2$$

This looks horrendous, until you recall that squaring and square-rooting are the inverse of each other, so they cancel each other out, giving:

$$d^2 = x^2 + y^2 + z^2$$

So, because we have used general terms here, we have derived the formula for Pythagoras' theorem in three dimensions. Many mathematicians and scientists consider more than three dimensions in their work and Pythagoras' theorem can be extended into as many dimensions as required. The latest work in quantum theory suggests that we live in an eleven-dimensional universe, but I think we'll leave it at three for this chapter!

Pythagoras – Man and Myth

Pythagoras may have been an outstanding mathematician and philosopher, but he wrote very little down. Our accounts of his life come from later sources, many of which take on a very mystical tone, imbuing Pythagoras with superpowers.

Sifting through this information for more probable details, we see that Pythagoras was a Greek who settled in southern Italy, then part of the Greek Empire. He founded a school of philosophy that was named after him which blended maths, science, religion and politics. It was split into two: the teachers, known as *mathematikoi*, and the listeners, known as *akoustikoi*. They had a few radical (for the time) ideas: that women should have equal standing with men, that a strict diet led to health in mind and body, and that numbers and shapes were divine. They also believed in numerology, the belief that the numbers associated with a person from their place or date of birth or name can influence their life.

The Pythagoreans' puritanical way of life and their political stance meant that they were often unpopular and had to be secretive. Pythagoras and his followers were burned out of town several times, fleeing to set up elsewhere.

Pythagoras' death is in as much doubt as the details of his life – some historians say he died when the temple he was in was barricaded and burned, others that he escaped and starved himself to death afterwards, others that he was captured and murdered because he refused to flee across a field of beans.

What is true is that Pythagoras and his school had a profound effect on the formation of Western philosophy.

Chapter 20

VOLUME

The volume of an object is how much space it takes up or encloses in three dimensions. Any solid that has straight edges is called a *polyhedron*. A cuboid is the simplest polyhedron to find the volume of as we multiply the length, width and height together. Another way of thinking about this is to say that it's the area of the face of the cuboid multiplied by the length.

Prisms and Spheres

We can extend this principle to any shape that has a constant cross section. Such shapes are known as *prisms*:

volume of prism = area of face × length

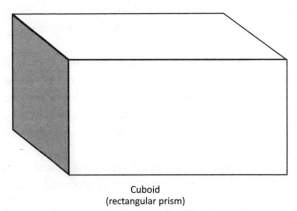

Cuboid
(rectangular prism)

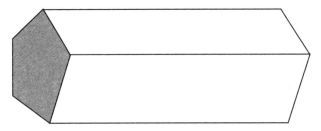

Pentagonal prism

For both these prisms, we would find the volume by working out the area of the grey face and multiplying by the length of the prism. We know all about area from the previous section, so the volume of prisms can usually be worked out without too much trouble.

When it comes to curvier objects, we refer back to Archimedes (see page 141). He worked out that the volume of a sphere is two-thirds of the volume of the cylinder that fits perfectly around it:

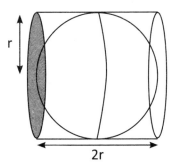

We can treat the cylinder as a prism. The area of the grey circular face is πr^2 and we need to multiply this by the length of the prism, which is 2r in this case:

$$\text{volume of cylinder} = \pi r^2 \times 2r$$
$$= 2\pi r^3$$

So to find the volume of the sphere, I need to multiply this by two-thirds:

$$\text{volume of sphere} = \tfrac{2}{3} \times 2\pi r^3$$
$$= \tfrac{4}{3} \times \pi r^3$$

Pyramids

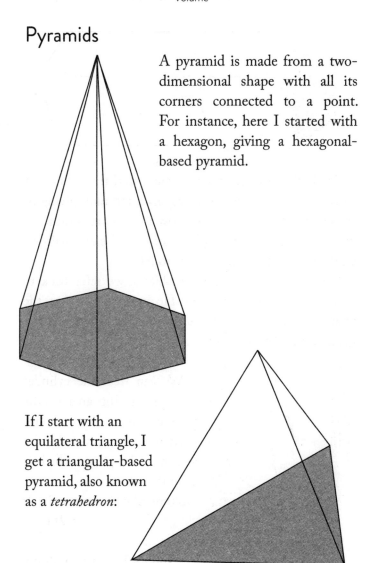

A pyramid is made from a two-dimensional shape with all its corners connected to a point. For instance, here I started with a hexagon, giving a hexagonal-based pyramid.

If I start with an equilateral triangle, I get a triangular-based pyramid, also known as a *tetrahedron*:

As with the sphere, there is a similar relationship between the volume of a pyramid and the volume of the prism that fits around it:

157

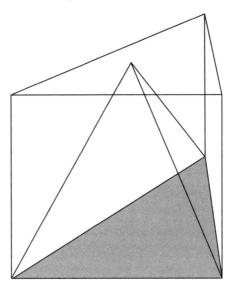

Aryabhata (see page 23) was amongst the first to find out that the volume of the pyramid is one-third the volume of the prism.

volume of pyramid = $\frac{1}{3}$ × area of face × length

Here, the face of the prism is the base of our pyramid and the length is really the height, so we can rewrite the formula as:

volume of pyramid = $\frac{1}{3}$ × base area × height

The Pyramids of Giza are square-based and the largest, known as the Great Pyramid, is truly colossal. The square base is 230 metres on each side, and when it was first built it was 147 metres tall. Using our formula:

volume of Great Pyramid = $\frac{1}{3}$ × *base area* × *height*

$\qquad\qquad\qquad\qquad\ = \frac{1}{3} \times 230^2 \times 147$

$\qquad\qquad\qquad\qquad\ = 2{,}592{,}100 \text{ m}^3$

To put this into context, the *Burj Khalifa*, the world's tallest building, has a volume of approximately 1.6 million m^3. Given that the Great Pyramid is 4,500 years older than the Burj and was made without the use of machines, the scale of it cannot fail to impress.

Eureka

When it comes to less regular solids, we have a couple of options. Either we approximate the volume of the solid using a combination of shapes that we do know about, or we can use *Archimedes' principle*. Legend has it that Archimedes was instructed to determine whether a crown that the King of Syracuse had commissioned contained all the gold that had been given for its construction. The king feared that some of the gold had been swapped for lead or silver, giving the jeweller a nice little bonus without reducing the mass of the crown.

The simplest way to do this would have been to melt the crown and compare its volume with the volume of the provided amount of gold. However, Archimedes was forbidden from harming the crown in any way.

As Archimedes lowered himself into a bath after a hard day's work on the problem, he realized that his body displaced the water. Thus, he could work out the volume of the crown by seeing how much water it displaced. Such was his joy that he ran down the street shouting '*Eureka!*' (Greek for 'I found it!'), neglecting, in his excitement, to get dressed again.

He ran the experiment the next day and the jeweller was found guilty.

5
STATISTICS

Chapter 21

AVERAGES

Statisticians are mathematicians that gather data, analyse them[†] to produce *statistics* that summarize them, and then make conclusions based on what the statistics tell them. Every aspect of our lives depends on statistics, from the size of our clothes, to the cost of our car insurance, to the medication we are prescribed when we are ill.

Whenever you are working out or analysing statistics, it's important to know whether the data come from a *population* or a *sample*.

A population is all the members of a particular group. For instance, if I were interested in the wingspans of adult Arctic terns, the population would be all adult Arctic terns. If I measured the wingspans of all adult Arctic terns, I could calculate the true statistics for the population.

In real life, it would not be possible to measure every single adult Arctic tern, so instead I would take a sample and hope that the sample statistics were similar to the population statistics. How I take my sample is important, because I want to avoid bias. For instance, if I take my sample from all the terns I manage to catch on one particular island, there's a possibility that all the terns in my sample may be related in some way, which may affect

---

[†] Data is the plural of datum. One datum, two data. You can't have a piece of data – that would be like having a piece of cakes.

their wingspan. There could be other groups on other islands that are significantly longer or shorter of wing.

There are various sampling methods that reflect how much time, effort and money you are prepared to spend. Sampling methods are very seldom reported along with the statistics presented to us daily. Shoddy sampling methods can be deliberately used to produce biased statistics, hence the famous quote attributed to the British prime minister Benjamin Disraeli (1804–81): 'There are three kinds of lies: lies, damned lies, and statistics.'

Some of the most commonly used statistics are *averages*. Averages usually work on the assumption that numerical data tends to clump in a central area – what statisticians call *central tendency*. As an example, most of your British female friends will be not too far from 164 cm or 5'4" tall, as this is the *mean* height.

The mean is the average we obtain when we add up all the data and divide by how many data there are – in effect we are sharing out the data evenly amongst every member of the sample. For example, I want to know the average height of my five-a-side team. First I measure the heights in centimetres:

167, 168, 175, 184, 191

These give a total of 167 + 168 + 175 + 184 + 191 = 885

I divide this by the number of data (5): 885 ÷ 5 = 177 cm

Notice that no one in the team is actually this tall, which is one reason why the mean may not be an appropriate value to use. At one time, the mean number of children per household in the UK was 2.4, which was widely mocked due to the impossibility of having anything other than a whole number of children. Numerical data may be divided into two kinds: *continuous* data that may take any value in a given range (like heights or weights) and *discrete* data that may only take certain values in the range, such as numbers of children or shoe sizes.

The *median* is the middle datum when they are arranged in numerical order – the same as lining up your sample in order of height or weight (or whatever) and picking the middle one. For my football team, the median is 175 cm. If we included the reserve, bringing the number of data to six, we would take the mean of the third and fourth data as the median.

If your data are not numerical, you can use the *mode*. This is not a measure of central tendency, but simply shows the most common or popular datum.

Averages by themselves can be very revealing, but more often than not we want some idea of how the data are spread out too. Are all the data near the average, or does the average just happen to be in the middle of two far-flung clusters of data? In the next section we'll look at ways of analysing this.

Getting It Wrong: Below Average

In 2012 the Chief Inspector of Schools in the UK stated that one in five primary school children were not reaching the national average in English, so standards needed to be raised.

A similar story is told of Dwight Eisenhower (1890– 1969), the US president who expressed shock that half of all US citizens were below average IQ.

A better understanding of averages would have helped both these gentlemen to raise their standards, too!

Chapter 22

MEASURES OF SPREAD

If I mark a set of maths exams, I'll usually calculate the mean score. This tells me how well the class performed on average and the students often want to know how they performed compared to the average. If the mean mark was 75%, it could mean that most students were either slightly below, at, or slightly above 75% – their marks are tightly clustered around 75%. This would imply that my students had similar ability and had responded to my teaching in the same way. Alternatively, the marks could be more spread out than this. I could have some students with low marks, balanced out by students with very high marks, to produce the same mean. How do I let my boss know the situation without making her sift through all the marks?

The answer is to give a measure of spread. There are several, of increasing complexity.

Range

The simplest is to work out the *range*: the highest score minus the lowest score. If the range of students' marks was 20, this implies that the marks in the exam went from about 65% up to 85% or thereabouts. The higher the range, the more the spread of the data.

As the range uses only the highest and lowest data, it can give a false idea of what happened. If one student had got 20%, but the rest were within 10% of the mean, the range would look big and wouldn't represent the data well. The 20% score may well be an *outlier* (see below for more details).

Interquartile Range

To avoid this problem, we can use the *interquartile range*, which tells us the spread of the middle 50% of the data. To do this, I need to work out the *quartiles*. The median is the middle datum, 50% of the way through the data, the central mark in the exam. If I look at the bottom half of the data and find the median of those, I get the datum that is 25% of the way through the data. This is called the *lower quartile*. A similar process with the top half of the data will give me the *upper quartile*, the datum that is 75% of the way through the data. The interquartile range is, therefore, the upper quartile minus the lower quartile.

The *box-and-whisker diagrams* below are for two groups of sixth-formers who took a maths exam:

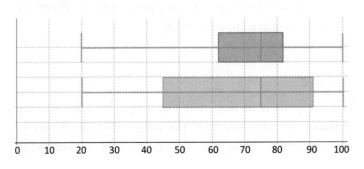

Each box-and-whisker diagram features five vertical lines. The two at the ends are the maximum and minimum data and these produce the whiskers. The three lines in the middle are the lower quartile, median and upper quartile. These make the box.

The box tells me about the middle 50% of the scores, while the whiskers show me the rest of the distribution. I can see from these diagrams that the range for both groups is 100 – 20 = 80%. The upper one has an interquartile range of 82 – 62 = 20%, so half the students had scores within 20% of each other.

The lower group of students have the same range (as shown by the whiskers) and the same median, but we can see from the box that their interquartile range is much larger (91 – 45 = 46%). From this, I can infer that the performance of the students in the second group was more varied than the first group, even if, on average, they did the same.

Standard Deviation

The standard deviation is a measure of how far the data are from the mean. It is not quite the average distance from the mean (which is called the *mean absolute deviation*), but it does have some very useful properties that we'll look at later on.

To calculate the standard deviation, we subtract the mean from each datum. This will give us negative values when the data are less than the mean, but we are only interested in the distance from the mean, not whether it was positive or negative. To sort this out we square all these values, as the negative values will become positive when squared (see page 63).

We add together all the squared values, divide by the number of data, and then take the square root to compensate for squaring earlier. This gives us the standard deviation. Here's an example:

Algebra test marks   Mean mark = 57.45
                           Number of students = 11

Mark	Mark - Mean	(Mark - Mean) squared
74	16.55	273.9025
44	-13.45	180.9025
45	-12.45	155.0025
42	-15.45	238.7025
45	-12.45	155.0025
76	18.55	344.1025
79	21.55	464.4025
40	-17.45	304.5025
38	-19.45	378.3025
83	25.55	652.8025
66	8.55	73.1025
	Total:	3220.7275

Standard deviation =

$$\sqrt{\frac{3220.7275}{11}}$$

= 17.11 (2dp)

This means that the scores of my students are quite spread out, not very consistent. A different group could have the same mean and range, but a smaller standard deviation, which would mean that the marks were closer to the mean on average.

Outliers

An outlier is a datum that doesn't fit with the rest, seeming to be much too low or high. There are a few arbitrary criteria for outliers – a datum is considered an outlier if it is:

greater than two standard deviations above the mean
less than two standard deviations below the mean
greater than 1.5 interquartile ranges above the upper quartile
or less than 1.5 interquartile ranges below the lower quartile

As an example, the mean height of an Austrian woman in 2001 was 167.6 cm, with a standard deviation of 5.6 cm. This gives an upper outlier of 167.6 + (2 × 5.6) = 178.8 cm, so any Austrian women over this height would be unusually tall. The lower limit is 167.6 − (2 × 5.6) = 156.4 cm, so any Austrian women under this would be considered unusually short.

When scientists collect data from observations and experiments, they have to be very careful with outliers. Is the outlier genuine so that it should be kept in the data set, or is it an error that should be deleted? Scientists often repeat experiments many times as this decreases the effect of an outlier on any statistics they produce.

In the 1970s NASA aircraft took regular measurements of the amount of ozone in the upper atmosphere. Flights over Antarctica often produced very low readings that the analysis software eliminated as outliers. It wasn't until ten years later that scientists working in Antarctica discovered the hole in the ozone layer. The ozone layer prevents most of the harmful ultraviolet radiation emitted by the sun from reaching the surface, so is crucial to life on Earth. Thankfully, thanks to a ban on ozone-destroying CFCs, the ozone layer is on the mend, but it will still take a few decades for the hole to be completely filled.

An outlier could be genuine or an error. Either way, they need very careful investigation.

Chapter 23

THE NORMAL DISTRIBUTION

Often, when you gather data, you find that a plot of the results gives a bell-shaped curve. For instance, if I measured the mass of the apples from my apple tree, I might get something like this:

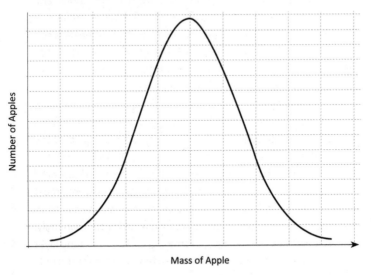

This is as we'd expect – most of the apples are close to a central value (the mean) and the further we go away from the mean, the lower the number of apples we find.

This curve is seen so often that it is tempting to think that this is the reason it is known as the *normal distribution*. The true reason is that statisticians use a normalized version – which just means it has a mean of zero and a standard deviation of one – to work out statistics about their data. This can then be used to work out what percentage of the population is greater than a given amount. This is the reason that the standard deviation is such a common measure of spread.

In the normal distribution, 68% of the data are within one standard deviation of the mean, 95% within two and 99.7% within three. When the scientists at the Large Hadron Collider announced the discovery of the Higgs boson in 2013, they talked about 'five sigma'. Sigma is the Greek letter used to represent standard deviation. Five sigma meant that the chance of their data happening by chance, without any new particles being involved, was five standard deviations away from the mean, giving a probability of about 0.0000003.

The normal distribution helps us in all sorts of ways. *Anthropometry* is the science of measuring the human body, and designers making everything from clothes, furniture and technology to trains, planes and buildings will use this data to decide what size everything needs to be. As these human measurements are normally distributed, they can work out what percentage of the population will fit into a medium-sized T-shirt or through a metre-square access hatch.

IQ – How Do Mathematicians Rank?

IQ, or *intelligence quotient*, is a somewhat controversial measure of intelligence. IQ is very hard to measure accurately and certainly can't be done with a ten-minute online test. IQs for a population are normally distributed, with a mean of 100 ('average intelligence') and a standard deviation of 15. Mensa, the high IQ society, bills itself as being for people in the top two per cent intelligence-wise, which corresponds to about 135.

In 1926 the American psychologist Catharine Morris Cox (1890–1984) published a very interesting study, part of which was to estimate the IQ of various eminent historical 'geniuses'.

Top of her list is the German polymath Johann von Goethe (1749–1832), a prodigious contributor to philosophy, politics, science and literature, with an estimated IQ of 188. Leibniz (see page 114) gets silver with 183, and the Frenchman Pierre-Simon Laplace (1749–1827) and Isaac Newton (see page 78) tie for bronze with 168.

Since then, the Australian mathematician Terence Tao (b. 1975), a Fields Medal winner, has an IQ rumoured to be over 200.

If you are feeling somewhat humbled, it may comfort you to know that human IQs seem to be drifting upwards as time goes by.

Chapter 24

CORRELATION

As well as trying to summarize data, we can also use statistics to look for relationships between data. If I gather two types of data from each member of a population or sample, I can plot these against each other on a *scatter graph*. Here is the scatter graph for the heights and weights of people in a gym:

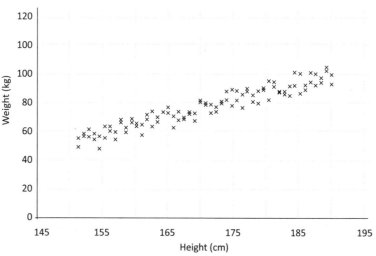

We can see that there is a general trend – the taller the person, the more they weigh. This is what we'd expect – there's more of a tall person than a small person, though this is offset by people's different builds and body shapes. Statisticians would call this a strong positive correlation.

Strong, because the points on the graph are close to a straight line, and positive because the two variables increase with each other.

This is an example of a weak negative correlation:

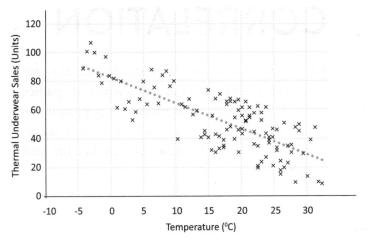

This correlation is negative because the graph shows that sales of thermal undies go down as the temperature increases. It is weak because the points, while they do indicate a trend, do not all lie close to the line shown. This line is called the *line of best fit* and is the line that passes as closely as possible to as many points as possible.

The Englishman Francis Galton (1822–1911) not only came up with standard deviation but was also one of the first to consider correlation and attempt to measure it. He was obsessed with measuring things and gathering data, and was particularly interested in anthropometry as he was a keen eugenicist – he believed that one should selectively breed humans to improve health and intelligence and to eradicate disabilities and other 'undesirable' traits. His work was picked up by his countryman

Karl Pearson (1857–1936), who found a mathematical way of measuring correlation (called *Pearson's product–moment correlation coefficient*) and also of drawing the perfect line of best fit. The PMCC varies from –1, for a perfect negative correlation, through 0 (implying no correlation at all) to 1 for a perfect positive correlation.

Pearson's coefficient only works for scatter graphs that produce a linear, straight-line relationship. Charles Spearman (1863–1945) got around this by ordering the data and using the rank of the data, rather than the actual number itself, to search for correlation. Hence, *Spearman's rank correlation coefficient* was born and no GCSE Geography project has been the same since. The diagrams below show a scatter graph of some data that clearly have a relationship, but it is not linear. If you rank the data and plot the ranks, you get the much more linear relationship shown on the bottom diagram.

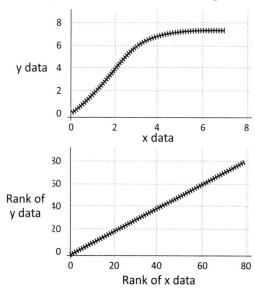

Correlation Is Not Causation

Just because things are correlated does not mean that one causes the other. Taller people weigh more, but putting on weight does not make you taller. There is a correlation between ice cream sales and drownings, but ice creams do not make people drown – hot weather makes more people want ice cream and want to go swimming, sadly increasing the number of drownings.

There is an excellent website by the American Tyler Vigen (tylervigen.com) that collates all kinds of spurious correlations, as he calls them. One of my favourites is:

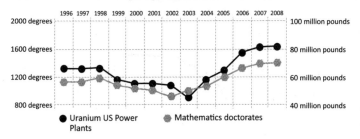

Mathematics doctorates awarded correlates with Uranium stored at US nuclear power plants

This is a great example that appears to show a correlation, but obviously putting uranium into US power stations will not cause maths PhDs to spring into existence.

6
PROBABILITY

Chapter 25

CHANCE

We weigh up chances all the time, knowingly or otherwise. Almost any activity entails some degree of risk, and risk is just a more sinister way of saying chance. Games of chance are stupendously popular. Fifty-five per cent of British people gamble, whether that is laying bets in a betting shop, playing the lottery, online poker or feeding a fruit machine.

Probability

Mathematicians talk about *outcomes* which *are the result of something happening*. For example, if I rolled a normal die[†], there are six possible outcomes. If I roll two dice and add the score, there are eleven possible outcomes (two to twelve).

Every outcome has a certain probability. We measure probability as an amount between zero (impossible) and one (certain). This means probabilities can be shown as fractions, decimals or percentages (see page 69). Anything with a probability of more than a half we could describe as likely or probable. Under a half, we say unlikely or improbable. Exactly a half is a fifty-fifty or even chance.

The outcome (or set of outcomes) you want to work

[†] One die. Two or more dice. I'm old-skool like that.

out the probability for is called an event. For instance, if I roll that die again, rolling a six would be an event. Some events are *mutually exclusive*, which means that they can't both happen at the same time. For instance, I can't pick a diamond and a heart if I choose one card at random from a pack of cards. These are mutually exclusive. I can, however, pick a diamond and a king. These are not mutually exclusive as there is a king of diamonds.

We often make some assumptions when we do probability. For instance, we usually assume that the dice we roll are *fair* or *unbiased*, which means each outcome has an equal chance of happening. We also assume that a chain of events are *independent*. If I roll a die and get a six, it doesn't change the probability of any outcomes the next time I roll it.

The magic formula for probability is:

$$P \ (\textit{Event happens}) = \frac{\textit{number of successful outcomes}}{\textit{total number of outcomes}}$$

The P() bit is shorthand for *probability of*. To find the probability of rolling a square number on a normal die, I see there are six possible outcomes (1, 2, 3, 4, 5, 6), of which two (1 and 4) are square numbers - the successful outcomes. I can write this as:

$$P(\textit{rolling a square number}) = \frac{2}{6} = \frac{1}{3}$$

I've cancelled the fraction down to its simplest form at the end. Now that I know the probability of rolling a square number, I can work out the probability of *not* rolling a square number by exploiting the fact that the two probabilities must add up to one.

$$P \ (\textit{not rolling a square number}) = 1 - P \ (\textit{rolling a square number})$$

$$= 1 - \frac{1}{3}$$

$$= \frac{2}{3}$$

In the 1650s, the mathematicians Pierre de Fermat (see page 150) and Blaise Pascal (see page 79) started to consider chance from a mathematical standpoint. Fermat had been asked by a professional gambler which was the surer bet:
• rolling at least one six in four rolls of a die
• rolling at least one double-six in 24 rolls of two dice

Instinctively, it feels like the latter should be the most likely as you get so many more rolls than the first bet. Let's take a look at each of these in turn.

At least one six in four rolls of a die
When I repeat an event, the number of possible outcomes of each event gets multiplied. Flipping a coin has two outcomes, but doing it twice has four outcomes: heads-heads, heads-tails, tails-heads, tails-tails. If I flipped it three times, there would be $2 \times 2 \times 2 = 8$ possible outcomes. This tells me that rolling a die four times will have $6 \times 6 \times 6 \times 6 = 1296$ possible outcomes. But how many of them count as successful? I can be successful with one, two, three or four sixes, and there are many different combinations to work out.

Something really handy to spot is that 'at least one' means the same as 'not zero' in this context. We saw above that the probability of something not happening is one minus the probability of it happening. So:

P(rolling at least one six = P (*not rolling zero sixes in four rolls*)
in four rolls) = 1 – P(*rolling zero sixes in four rolls*)

There are five ways to roll a die and not get a six, so there must be $5 \times 5 \times 5 \times 5 = 625$ successful outcomes for getting no sixes when I roll four times:

1 – P (rolling zero sixes in four rolls) $= 1 - \frac{625}{1296}$

$= \frac{671}{1296}$

This gives the probability of getting at least one six in four rolls as 51.8%.

Rolling at least one double-six in 24 rolls of two dice
Rolling two dice and adding the scores is a mainstay of many games. It's not immediately obvious what the probability of rolling a given total is, so making a table of all the outcomes can help:

+	1	2	3	4	5	6
1	2	3	4	5	6	7
2	3	4	5	6	7	8
3	4	5	6	7	8	9
4	5	6	7	8	9	10
5	6	7	8	9	10	11
6	7	8	9	10	11	12

This table is known as a *probability space diagram*. We can see that there are 36 equally likely outcomes. Seven is the most likely outcome as it has the most combinations of scores that lead to it. Double-six, or twelve, has only one, so the probability of it happening is one-36th.

Much like the previous example, 'at least one' means 'not zero', so I can look at the probability of rolling no double-sixes in 24 throws. If the probability of rolling a double-six is $\frac{1}{36}$, then the probability of not rolling a double-six must be $\frac{35}{36}$, as $\frac{1}{36} + \frac{35}{36} = 1$.

P (*rolling at least*   = 1 – P (*rolling zero double-sixes in 24 throws*)
one double-six   = $1 - \left(\frac{35}{36}\right)^{24}$
in 24 throws)

                   = 0.491 (to three decimal places)

So the probability of rolling at least one double-six in 24 throws is 49.1%. The gambler would be better off taking the first bet, but not by much. This counter-intuitive result helped the gambler to understand his lack of success at the dice table.

What a Coincidence

The *birthday problem* was originally posed by the Ukrainian engineer Richard von Mises (1883–1953). Probability problems often yield counter-intuitive results, and this one is no exception. If you went to a café, it seems unlikely that any two people there would share a birthday unless hundreds of people were present, as there are 365 days in a year. However, as we are looking at so many different possible pairs of people, the chance of any two of them sharing a birthday actually reaches 50% if there are only 23 people present. It rises to 99.9% if you have 70 people.

Chapter 26

COMBINATIONS AND PERMUTATIONS

I have a class of twenty-four equally brilliant maths students. I need to pick four of them for a maths competition and it is so hard to choose between them that I decide to do it at random. How many different possible teams are there?

This depends on whether the order of selection matters. For instance, if the first person I pick is going to be the captain, the second wields the calculator, the third writes things down and the fourth makes the tea, then the order matters.

If I draw names out of a hat, I have 24 possibilities for the first member of the team, 23 for the second, and so on, giving me:

$$24 \times 23 \times 22 \times 21 = 255024$$

So I have 255,024 *permutations* of my team. This is the same as:

$$24 \times 23 \times 22 \times 21 = \frac{24 \times 23 \times 22 \times 21 \times 20 \times 19 \times 18 \times 17 \times 16 \times 15 \times 14 \times 13 \times 12 \times 11 \times 10 \times 9 \times 8 \times 7 \times 6 \times 5 \times 4 \times 3 \times 2 \times 1}{20 \times 19 \times 18 \times 17 \times 16 \times 15 \times 14 \times 13 \times 12 \times 11 \times 10 \times 9 \times 8 \times 7 \times 6 \times 5 \times 4 \times 3 \times 2 \times 1}$$

Why write it like that? Well, if I use factorial notation (see page 121) I can reduce this to something easier to enter into my calculator:

$$24 \times 23 \times 22 \times 21 = \frac{24!}{20!}$$

In general, if you are choosing k things out of n in total:

$$\text{number of permutations} = \frac{n!}{(n-k)!}$$

So, if I were choosing a team of six instead, I would have k = 6 with n = 24:

$$\text{number of possible teams} = \frac{24!}{(24-6)!} = \frac{24!}{18!} = 96909120$$

A lot of the four-student teams I could draw from the hat would contain the same people, just in a different order of selection. If the order doesn't matter, this effectively makes them the same team – selecting Amy, Billy, Cara and Dan in that order gives the same team as selecting Dan, Cara, Billy and then Amy. I can arrange four people in $4 \times 3 \times 2 \times 1 = 4!$ different ways, so I need to divide the number of permutations by this to get the number of possible *combinations*:

$$\text{number of combinations} = \frac{24!}{20!4!} = 10626$$

And again, in general, if you are choosing k things out of n in total where order doesn't matter:

$$\text{number of combinations} = \frac{n!}{(n-k)!k!}$$

So, if I were choosing a team of six where order doesn't matter, I would have:

$$\text{number of possible teams} = \frac{24!}{(24-6)!6!} = \frac{24!}{18!6!} = 134596$$

If you've followed the above closely, you'll be aware that combination locks are misnamed – as order matters, they should technically be called permutation locks.

Permutations and combinations can help us solve probability problems. The UK national lottery requires you to choose six numbers from one to forty-nine. To win the top prize, you must match all six numbers drawn at random. Order doesn't matter in the lottery, so combinations are the key.

$$\text{number of possible 6-number combinations} = \frac{49!}{(49-6)!6!} = \frac{49!}{43!6!} = 13983816$$

So there are just under 14 million possible combinations, making your chance of winning the top prize one-14-millionth. Not huge!

Chapter 27

RELATIVE FREQUENCY

In the examples earlier, we were able to work out the total number of possible outcomes and so calculate our probabilities using only theory. In many circumstances, however, it's not possible to do this. If you asked me what the probability of me having a cup of coffee today is, I could tell you that it is very high, and even estimate a figure, but I couldn't work it out mathematically without gathering some data first.

I could keep a coffee diary for a week and use that to work out a probability. Let's say in the first week I have coffee on five out of the seven days. We could then say, based on this evidence, that the probability of me having a coffee on any given day is $\frac{5}{7}$. Mathematicians call this a *relative frequency* to show that it is not a theoretically derived probability. We have to assume that having coffee on a day is independent – that I'm not more or less likely to have a coffee if I had one the day before.

The next week I have coffee every day. My relative frequency is now:

$$\text{relative frequency} = \frac{5+7}{7+7} = \frac{12}{14} = \frac{6}{7}$$

The general idea is that the more time that passes, the more accurately the relative frequency represents the probability that I will have a coffee on any given day.

Why is this useful? Well, when bookies set odds and gamblers make bets, they will generally be looking at recent past performance (*form*) to inform their decisions. Insurance companies use a similar process to pigeonhole their customers to ascertain the risk of insuring them and set the premium appropriately. The concept behind a no-claims bonus is that the longer you go without making a claim, the lower the relative frequency of you making a claim becomes, so the less risk you pose to the insurer, so the lower the premium they offer you.

Probability Fallacies

If I flip a coin and get heads eight times in a row, a lot of people feel that the universe is now out of balance somehow and start to believe that getting a tails becomes more likely as a result, whereas in actual fact the chances of tails are still 50%. Although getting heads eight times in a row is unlikely (about 0.4%), it has the same probability as any other combination of eight flips.

This mistake is known as the *gambler's fallacy*. A famous incident occurred at the Monte Carlo Casino in 1913. One of the roulette wheels landed on black twenty-three times in succession, purely by chance. Word quickly got around while it was happening, with people betting large sums on the next spin ending up on red in the belief that this was more likely to happen.

Some people make the same mistake when having

children – assuming they are more likely to have a girl if they already have several boys or vice versa.

The *prosecutor's fallacy* is the mistaken belief that in a court case, the probability of a claimed event occurring is the same as the probability of the accused being guilty or innocent, as appropriate. Sally Clark (1964–2007) was a British woman convicted of murdering her two children, who both actually died from sudden infant death syndrome. The odds of this having occurred were mistakenly calculated as one in 73 million, assuming that SIDS deaths are independent, which cannot be assumed for siblings as there may be an underlying genetic condition. She served three years of her sentence before she was acquitted and many other similar cases were reviewed.

Getting your statistics wrong can have very serious consequences indeed.

AFTERWORD

If you've read the book through to this point, you have enjoyed a mighty six-course feast. Hopefully, this has made you realize that mathematics is something accessible to everyone, at every level.

There are a great many textbooks that you could use to study further if you so wanted, and the internet is awash with many excellent free resources too. If you don't want to push your knowledge of mathematics so much as find out more about the backstories, your library, bookshop or search engine can put you on the right track.

If you have enjoyed these bite-sized chunks, please make mathematics part of your regular diet. Mathematics really does make the world go round and the more mathematically literate we can make the world, the better it will be.

I wrote at the beginning of the book that mathematical anxiety is infectious, but so is mathematical confidence. If you've gained confidence, share it with those around you, let them know that your understanding of maths can improve if you take the time to read and learn.

Don't stop here with this book – carry on the journey and explore all the tasty mathematical dessert that's out there. Become a gourmet – you don't need to understand exactly how a meal is cooked to appreciate the skill that has gone into it and to enjoy the final product.

Bon appétit!

INDEX